The Politics of
Nuclear Waste

Pergamon Titles of Related Interest

Auer ENERGY AND THE DEVELOPING NATIONS
DeVolpi PROLIFERATION, PLUTONIUM AND POLICY: Institutional and
  Technological Impediments to Nuclear Weapons Propagation
Gabor BEYOND THE AGE OF WASTE: Second Edition
Jackson NUCLEAR WASTE MANAGEMENT: The Ocean Alternative
Mossavar-Rahmani ENERGY POLICY IN IRAN
Neff THE SOCIAL COSTS OF SOLAR ENERGY
Stewart TRANSITIONAL ENERGY POLICY 1980-2030: Alternative
  Nuclear Technologies
Vogt ENERGY CONSERVATION AND USE OF RENEWABLE ENERGIES IN
  THE BIOINDUSTRIES
Williams/Deese NUCLEAR NONPROLIFERATION: The Spent Fuel Problem

Related Journals*

ENERGY
ENERGY CONVERSION AND MANAGEMENT
NUCLEAR AND CHEMICAL WASTE MANAGEMENT
PROGRESS IN NUCLEAR ENERGY
SUNWORLD

*Free specimen copies available upon request.

PERGAMON
POLICY
STUDIES
ON ENERGY

The Politics of
Nuclear Waste

Edited by
E. William Colglazier, Jr.

An Aspen Institute Book

Pergamon Press
NEW YORK • OXFORD • TORONTO • SYDNEY • PARIS • FRANKFURT

820778

Pergamon Press Offices:

U.S.A.	Pergamon Press Inc., Maxwell House, Fairview Park, Elmsford, New York 10523, U.S.A.
U.K.	Pergamon Press Ltd., Headington Hill Hall, Oxford OX3 OBW, England
CANADA	Pergamon Press Canada Ltd., Suite 104, 150 Consumers Road, Willowdale, Ontario M2J 1P9, Canada
AUSTRALIA	Pergamon Press (Aust.) Pty. Ltd., P.O. Box 544, Potts Point, NSW 2011, Australia
FRANCE	Pergamon Press SARL, 24 rue des Ecoles, 75240 Paris, Cedex 05, France
FEDERAL REPUBLIC OF GERMANY	Pergamon Press GmbH, Hammerweg 6 6242 Kronberg/Taunus, Federal Republic of Germany

Library of Congress Cataloging in Publication Data
Main entry under title:

The Politics of nuclear waste.

    (Pergamon policy studies on energy)
    Includes index.
    1. Radioactive waste disposal--United States--
Addresses, essays, lectures. 2. Atomic energy
policy--United States--Addresses, essays, lectures.
I. Colglazier, E. William (Elmer William)
II. Series.
TD898.P64   1981   621.48'38   81-10589
ISBN 0-08-026323-2   AACR2

Printed in the United States of America

Contents

Appendices

Foreword

Honorable Richard W. Riley
Governor of South Carolina
Chairman of the State Planning Council on
Radioactive Waste Management

NUCLEAR WASTE AND GOVERNANCE

Two years ago I wrote in the *Washington Post* that my decision to ban the disposal in South Carolina of low-level nuclear wastes from the crippled reactor at Three Mile Island was based on the belief that my state could "no longer be the path of least resistance in seeking a national answer to nuclear waste disposal." Let me describe the circumstances which led me to this conclusion and some of the things that have occurred since which are moving us toward a "national answer."

In 1967, the South Carolina General Assembly passed *The Atomic Energy and Radiation Control Act*. With this act, South Carolina became an "Agreement State" with the Nuclear Regulatory Commission. Simply put, this allowed South Carolina to assume direct responsibility for and control of the licensing and permitting of radioactive material use in the state. State responsibility does not include, however, operations of the federal government such as the Savannah River Plant. Nor does it include the on-site operations of nuclear power plants; South Carolina has four reactors in operation now, one almost ready to begin operation and four more planned or under construction.

In 1969, a licensing application was received by the state for the establishment of a commercially operated low-level nuclear waste burial facility. It took two years of scrutiny and further study before the license was granted in 1971. In retrospect, it was that two years of careful consideration which enabled South Carolina to have a low-level radioactive waste disposal facility that has not required closure due to the operational problems other states have experienced.

Shortly after I took office in January of 1979, it was brought to my attention that what had been created in 1971 as a regional facility—one of six low-level radioactive waste burial sites in the United States—was fast on its way to becoming one of only two, or maybe the only one, in the country. More than three million cubic feet of low-level radioactive waste were placed in commercial burial grounds in 1978. Because of the proximity to the majority of generators of such waste, nearly 80 percent of it came to South Carolina for disposal.

Shipments were consistently arriving at the site that did not comply with applicable laws or regulations. Many discrepancies involved radiation readings in excess of allowable limits. The states in which the wastes originated were assuming little, if any, responsibility for the packaging or transportation methods practiced by waste generators. I asked our Department of Health and Environmental Control to begin inspection of every shipment of the waste when it arrived at the burial site.

In six months of intensive inspection, we demonstrated not only that shipments were being made improperly, but that our reports of these discrepancies were not being dealt with appropriately by federal authorities. I joined the governors of the states where two other sites are located, Governor Dixy Lee Ray of Washington and Governor Robert List of Nevada, in calling for the Nuclear Regulatory Commission (NRC) and the U.S. Department of Transportation to do their jobs by enforcing their regulations and sanctioning violators. A regulatory process that is not enforced is nothing more than sham, and that is what existed in the summer of 1979.

We also turned to state law to sanction the violators and found that no civil penalties were provided. The extent to which we could show, however, that a violation, or its repeated occurrence, constituted a threat to the public health, we were able to issue orders prohibiting further shipments from that generator until such time as they were able to clearly demonstrate that disregard for law and regulation would not recur.

The federal authorities have since demonstrated a better ability to enforce laws and regulations. In addition, by the end of the 1980 legislative session, South Carolina had passed amendments to the South Carolina Atomic Energy and Radiation Control Act providing authority for civil penalties to punish those who do not ship waste in our state exactly by the letter of the law. In addition, every state was notified that we were reducing the amount of low-level waste allowed into our site each year. We were receiving as much as 250,000 cubic feet a month in 1979, when the South Carolina site had become, through attrition, the national disposal facility for low-level nuclear waste. In October 1979, South Carolina instituted a license condition that established volume limits on the amount of wastes that could be accepted at the facility near Barnwell. By late 1981, no more than 100,000 cubic feet per month will be allowed.

Continued monitoring, reduced volume, sanction of violators, prohibiting access to our disposal site by those who repeatedly violate law and regulation— those were the things we could do as a state to assure our citizens, with confidence, that continued operation of the site would be proper.

What remained to be done was to get other states to accept responsibility for the low-level radioactive wastes generated within their borders and to secure the creation of new disposal sites located regionally across the country.

I have advocated since my first days in office that Congress take action to mandate that the individual states be responsible for the management and disposal of low-level nuclear waste generated within their borders. This proposal would remove the remaining barrier to the creation of new disposal sites, that is, the perception by most states that this is a federal responsibility and that the federal government would take their wastes to new federal sites or existing federal sites now used almost exclusively for wastes from federal research and defense activities. In addition to assigning this responsibility to the states, it was necessary that states be authorized to enter into interstate compacts for managing such wastes and that the compacts have the right to exclude low-level nuclear wastes generated outside those states comprising the compact. This authority would be the real impetus for the development of new disposal capacity by allowing regions with sites to determine from whom receipt of wastes would be allowed. It is clear that when management and disposal become a state responsibility, rather than that of the federal government, states must either create their own sites or join with other states to do so.

These basic principles of a low-level waste management system—state responsibility, a regional approach to the development and maintenance of disposal facilities, and the right to exclude wastes from outside a regional compact—were all endorsed by the State Planning Council on Radioactive Waste Management, the National Governors' Association, the National Conference of State Legislators, and many other national organizations.

In December 1980, Congress passed legislation establishing these principles as the basis for federal policy in low-level waste management, and this legislation was signed into law by President Carter. With the Low-Level Waste Policy Act of 1980, Congress has created the framework within which all states must assume their proper responsibility for the disposal of low-level radioactive waste.

As soon as I became involved in the matter of low-level nuclear waste, I was visited by some federal officials who wanted to talk about South Carolina having another type of nuclear waste facility—a storage site for spent nuclear power reactor fuel. My response was, and has consistently been, opposition to the idea of the federal government providing away-from-reactor (AFR) storage without a demonstrated commitment to the permanent disposal of high-level radioactive wastes. Even then, consideration of an AFR would have to be on a limited basis and according to a regional system.

There is basic law of political-physics, often overlooked in nuclear waste considerations, which says that waste stays where it is first put. What may be intended to be a *temporary* interim storage could, without a demonstrated commitment to permanent disposal of high-level wastes, become de facto permanent storage. One just need look at the more than 23 million gallons of high-level radioactive wastes that have been accumulating in temporary storage at the Savannah River Plant for almost three decades.

Lately, proposals have been brought forward to reinitiate the reprocessing of commercial spent nuclear fuel. One possible location for this reprocessing would be at a facility under construction in Barnwell County, South Carolina. There are certain attractive aspects to the reprocessing option; for example, the recovery of the energy value in the unused portion of the fuel and the production of a waste form that may be easier to dispose of. However, current reprocessing suffers from the same limitations as the AFR proposal. In the absence of a disposal capacity or a demonstrated commitment to develop one, the establishment of either an AFR or a reprocessing facility in Barnwell County will result in the accumulation in South Carolina of more nuclear waste with no place to go.

This problem of high-level radioactive waste disposal has been of central concern to the State Planning Council. For over three decades now, the federal government has talked about the need to develop a permanent disposal capacity for these wastes, but has made relatively little progress in achieving that object. Thus far, they have largely found out what not to do in locating a permanent repository for high-level wastes. While many of these lessons have been important learning experiences, they do not represent the kind of sustained, programmatic thrust which will lead to a solution to this problem.

The Council has recommended that the primary emphasis of the federal government's waste management efforts must remain centered on finding a solution to the disposal of high-level wastes. The Council has urged that this be done through a technically conservative, step-by-step program based on the investigation of multiple sites for geologic disposal.

The Council has also recommended that the federal government's search for permanent repositories be conducted through a process of consultation and concurrence with the states and tribes. For this process to be effective, it will require a real partnership among elected officials at all levels of government in this country—a partnership that rejects both the arbitrary imposition of federal will and arbitrary actions by states or tribes to block efforts to dispose safely of nuclear wastes.

What we are attempting to do in South Carolina, in the State Planning Council, and in states, tribes, and local communities throughout the country is to contend with the entry of the political process into nuclear waste decision making. Heretofore, such decisions were the sole purview of those trained in the mystical sciences that produced new elements of nature, and our faith in them was absolute. Such absolute faith has, however, been shaken and our attitudes toward such decision making have been drastically transformed.

Because of this transformation, we have gone beyond the question of "what will we do with this waste?" There are some who still say that public officials, like governors and state legislators, do not understand, much less know how to resolve, issues involving complex questions of nuclear physics and that the keystone to our democracy, the voter, simply cannot grasp many of the facts having a bearing on this question. Therefore, they contend that the decision on "what to do" should remain solely in the community of those who *do* understand.

This simply cannot be.

The public is aware and is concerned. It may be that, on occasion, their reactions are not equal to the risks. But the public's concerns are real nonetheless. And without a serious effort to rebuild public confidence, our ability to manage these wastes will be seriously threatened.

It has been said that we were lulled into a false sense of security regarding nuclear waste, its storage, and its disposal. The very manner in which past decisions were made simply did not provide the opportunity for adequate critical assessments that might have encouraged new knowledge and technology to solve the questions facing us today.

Therefore, if the questions are truly "Who will decide what will be done?" and "How will they decide?", those of us who are involved in these decisions must ensure that the demands of the political process are balanced with and informed by the appropriate scientific expertise. Within such a scientific/ political debate, all parties must be prepared to substantiate the basis for their recommendations. It will be this substantiation, presented in a spirit of openness, available for scrutiny and criticism, and this kind of substantiation alone, that will provide the *confidence* so necessary to nuclear waste management decisions.

Of highest importance, today, is not only *what* is to be done, but also *how* we decide it is to be done. A process of decision making must be established that will allow us to have confidence in the results of that process. There will be remaining uncertainties no matter what the decisions are. Only confidence in the process which leads to those decisions will enable us, as a society, to live with those remaining uncertainties.

Preface

E. William Colglazier, Jr., and Paul Doty

Radioactive wastes present society with a novel and difficult technological problem that will test anew the capabilities of our governance institutions. No matter what happens to nuclear power in the future, the extremely hazardous wastes from plutonium production for military purposes and from electricity generation by existing commercial reactors will have to be safely managed for centuries to come. In the United States, the current volume and radioactivity of high-level waste, which includes spent reactor fuel and fission product waste from reprocessing, exceeds 290,000 cubic meters and one billion curies, respectively. The generation of these high level wastes is continuing, with quantities increasing particularly rapidly for commercial spent fuel where the cumulative radioactivity content already surpasses that of military waste. (Definitions, current quantities, and projected quantities by waste type and institutional origin are given in the fact sheet in Appendix A.)

From interim storage and transport to permanent disposal, from reprocessed waste and spent fuel to low-level waste, all areas of radioactive waste management abound in contentious social, political, and institutional issues. After four decades into the nuclear era, many observers now have come to realize that "the resolution of institutional issues . . . is equally as important as the resolution of outstanding technical issues" in nuclear waste management, and that such resolution "may well be more difficult than finding solutions to remaining technical problems."[1] The origin of some of the disputes is as fundamental as the debates over the sharing of power in the federal system, the role of government in the commercial sector, and the extent of public participation in a military activity. Always lurking in the background is the potential linkage to the future of nuclear power, as well as the unsettling image of the awesome destructive power of nuclear weapons.

The complexity and interrelation of these issues is illustrated in the successive attempts to site a respository. The federal government has the responsibility for the eventual permanent disposal of all high-level radioactive waste in the United States, and the government's favored technical strategy is placement in mined geologic respositories deep underground.[2] Wide agreement has developed that the next step requires in-situ examination and experimentation in a variety of geologic formations. Yet, gaining access from states, even to characterize potential sites, has encountered many difficulties. Many states and local communities have declared outright bans on waste disposal or its transport within their borders. Even with the legal supremacy of the federal government, persistence of states in raising these roadblocks might nevertheless ensure that political considerations will dominate technical judgment.

The understandable reluctance of states and local communities to accept either technical or institutional assurances from the federal government on radioactive waste management is rooted in unpleasant historical experience. Federal agencies have at times been myopic in seeking technical fixes while ignoring legitimate social and political concerns. Most state and local elected officials have not been given much voice in the decision-making process, have feared the power of federal preemption, and have felt that the hazards may not have been fully explained to them or may not be sufficiently understood. The experiences in seeking sites in the early to mid 1970s in Kansas and Michigan did not inspire confidence in either the candor or the competence of federal officials.[3] It should have been clear from these incidents that "a federal agency disregards at its peril the potential power of state and local officials whose opinions reflect the consensus of their constituency on matters of health and safety."[4]

The recent experience in New Mexico is a reflection of these problems; it confirms that federal promises are still unreliable and that the institutional source of nuclear waste is more important politically than any physical characteristics. The proposed Waste Isolation Pilot Plant (WIPP) repository in southern New Mexico was originally authorized as a permanent disposal facility for defense transuranic waste, primarily to fulfill a promise made informally in 1970 to Idaho officials that waste temporarily stored at the National Engineering Laboratory would be moved in a decade. The intended recipient, New Mexico, was later told by the Secretary of Energy that it would effectively have a veto right over any waste disposal site within its borders. In February 1980, the state was told by the President that the proposed WIPP facility near Carlsbad would be cancelled while alternative sites were investigated nationally.[5] If a repository were to be built on the WIPP site, President Carter said that it would have to be licensed by the Nuclear Regulatory Commission (NRC) and, for economic and administrative efficiency, would receive both civilian and military wastes.

At nearly the same time as the President's message, however, Congress directed the reversal of all these executive branch commitments by statutorily requiring WIPP to be an unlicensed facility solely for military waste, including research on high–level waste and disposal of transuranic waste.[6] The conflict between the President and the Congress over WIPP was mushrooming into possibly another interminable battle over nuclear policy, such as that which occurred over the Clinch River Breeder Reactor. The Congress had nevertheless promised a consultation and cooperation agreement with New Mexico.

Yet, on January 23, 1981, three days after the new administration took office and without a signed agreement in place, New Mexico was perturbed to learn that the Department of Energy (DOE) was proceeding expeditiously with developing the nuclear waste burial site near Carlsbad and that the federal project managers "don't need anything else from the state, legally or officially."[7] *The Albuquerque Journal* rejoined in its editorial, "New Mexicans justifiably are skeptical of promises made by DOE officials."[8] The Attorney General of New Mexico brought suit against the federal government. The dispute was temporarily settled when DOE agreed to allow New Mexico the opportunity to review technical information after sinking the shaft and testing at depth and to seek resolution of other state concerns before the final commitment to construction of the permanent repository.[9] The Secretary of Energy retained the ultimate authority to make final decisions, but New Mexico received a judicially enforceable agreement and reserved its right to litigate in the future. Whether these changing tactics by the federal government will be successful in putting waste in the ground remains to be seen.

The institutional issues surrounding the interim storage of commercial spent fuel—who has responsibility, who will pay, and where will it reside—are no less contentious. After President Carter's indefinite deferral of commercial reprocessing in 1977, the administration proposed to Congress that the federal government assume responsibility for away-from-reactor (AFR) interim storage of utility spent fuel.[10] This proposal was made in the context of Department of Energy projections that some operating reactors would run out of the existing on-site storage capacity in the mid-1980s—regardless of decisions on commercial reprocessing—and that considerable additional capacity would be needed in the 1990s until a permanent repository opens.[11]

The Congress has yet to determine what role, if any, the federal government should play in alleviating this near-term problem faced by utilities. Clearly, the failure by society to deal with this problem could shut down operating reactors. The three states which have unused commercial reprocessing facilities are extremely reluctant to let the federal government use them as the first stage of a federal AFR system. As Governor Richard Riley has stated, "I would be adamantly opposed to an AFR in South Carolina in the absence of a commitment to a true plan for permanent disposal of nuclear waste. There is a basic law of nuclear waste often overlooked—all waste remains where it is first put— just look at the 23 million gallons of temporarily stored (military) waste at the Savannah River Plant."[12] The clout behind this state perspective has been enhanced by President Reagan's reversal of the promise of federal AFRs, as part of his commitment to reduce the role of government in commercial activities.[13]

For low-level waste, the governors of the three states with operating commercial dumps announced in 1979 that they would no longer assume the full burden of the nation's low-level waste disposal problems and called for the establishment of additional disposal sites on a regional basis. Their collaboration on the temporary closures of two of the sites and the announced volume reductions for the third have forced attention to this issue, and some progress has resulted. Other states now appear ready to accept responsibility for insuring that their own commercial low-level wastes be taken care of, and Congress has endorsed the concept of regional compacts that will have the authority to exclude wastes from non-member states.[14] But even with an institutional solution for this much simpler low-level waste problem now visible, much remains to be done before new disposal facilities are operational.

Because of growing concern about the volumes of waste and its safety in transport, more than half of the states and many localities have passed ordinances that restrict or prohibit the passage of radioactive waste shipments within their jurisdictions. Cities varying in size from New York City to Missoula, Montana, have adopted this course. Shippers have complained that they are being forced to take circuitous routes over second-class roads and that interstate commerce is being unduly obstructed. In response, the Department of Transportation began in 1978 a rule-making process to define federal regulations for a uniform highway routing system that would preempt restrictive local ordinances. Although the final rules do not go into effect until February 1, 1982, many states are unhappy that reasonable prenotification requirements for high level or large volume shipments may also be overridden, and that the approved processes for designating alternate routing may not be ready.

The social, political, and institutional dilemmas surrounding radioactive waste do not end with repository siting, interim spent fuel storage, transportation, and low-level waste. In particular, large volumes of military waste continued to be stored in a less than satisfactory interim manner, abandoned mill-tailing piles are spread throughout the West,[15] and numerous military and commercial facilities will have to be decontaminated and decommissioned at some future date. While constructive steps are being made towards building a national consensus in some areas of nuclear waste management, the skeptic might claim that a stable governance framework is further away than ever.

ORIGIN OF THE BOOK

Two years ago the Aspen Institute for Humanistic Studies adopted the issue of "governance" as the integrating theme of its activities. The generic question posed was how well suited are our institutions, public and private, to serve fundamental values of society such as freedom, order, and justice.

In November of 1979, the Program in Science, Technology and Humanism and the Energy Committee of the Aspen Institute organized a conference on "resolving the social, political and institutional conflicts over the permanent siting of radioactive wastes." The two-day conference, held at the John F. Kennedy School of Government, Harvard University, aimed at exploring the governance issues in radioactive waste management. It was not designed to reach a consensus or make recommendations, but rather to allow fifty participants from government, industry, utilities, environmental groups, the legal profession, states, the media, academia, labor and civic organizations to educate and lobby each other. By clarifying divergent views on the nuclear waste issue, it was hoped that the essential problems would become more sharply defined and the principal options elucidated. In so doing, each observer might then gain insight into the debate and judge for himself or herself the possible elements of consensus and the best course of action.

This volume is an outgrowth of that Aspen Institute conference in November 1979. Several important events for radioactive waste management occurred subsequently in 1980: the announcement of a comprehensive policy by President Carter; the creation of the State Planning Council on Radioactive Waste Management, composed of state and local elected officials and federal agency heads; the passage of bills by both the Senate and House whose differences were almost resolved in the waning days of the 96th Congress; and the election of President Reagan. The first year of a new administration is a good time to reflect on all that has passed, particularly in the last four years, and to see if further progress is possible or destined to remain elusive. The following chapters and appendicies, some having begun as commissioned papers for the Aspen Institute conference, provide a comprehensive and up-to-date overview of the governance issues connected with radioactive waste management as well as a sampling of the diverse views of the interested parties. In offering a forum for this frank interchange of opinions and a personal estimate of the prospects for consensus, we implicitly hope that our governance institutions will soon be able to provide the stable organizational framework needed for constructing a safe and societally acceptable solution to the nuclear waste management problem.

PERSPECTIVES FROM THE ASPEN INSTITUTE CONFERENCE

A brief recapitulation of the various perspectives on high-level waste disposal presented at the Aspen Institute conference may serve as a useful prologue to this volume. One easy way of describing the diversity of views is in terms of four stereotypical positions, which can be labelled "some grassroots environmentalists," "the IRG purists," "the Washington pragmatists," and "some friends of nuclear power," where IRG refers to the point of view advocated by the Interagency Review Group during the Carter administration. These categories are truly stereotypes and are not meant to apply precisely to particular individuals or groups. For example, an environmentalist might find that his views have something in common with several or all of the four categories. Nor is there an implication that the middle two positions represent the best framework for consensus or the best public policy. But by surveying all four positions, one can be gently sensitized to this lively debate in preparation for the details to come.

Some grassroots environmentalists believe that the radioactive waste issue is the cutting edge for grassroots antinuclear organization because nobody is pro-waste. In their opinion, an increasingly informed citizenry will eventually stymie the waste-siting process and ultimately demand an end to the production of more waste, thereby stopping nuclear power. They are not particularly interested in consensus-building efforts because they would rather stake their case on participatory democracy working its will. Licensing of waste facilities by the Nuclear Regulatory Commission is strongly favored since it allows more public intervention. Moreover, this group feels that heightened public awareness will lead to more safety when wastes are eventually disposed of, that is, after nuclear power is stopped.

The IRG purists are faithful adherents and strict constructionists of the Interagency Review Group document produced in March 1979 for President Carter.[16] This view includes some representatives of states which are either potential hosts for a permanent repository or current hosts for temporarily stored wastes. States in this category include New Mexico with the planned WIPP facility, New York with the liquid high-level waste stored at the closed West Valley reprocessing plant, Idaho with transuranic waste stored at the National Engineering Laboratory, and many others whose geologic formations are now being actively investigated by government scientists.

Through legislation, the IRG purists want the states to be given a strong "consultation and concurrence" right over repository siting—the phrase invented by the IRG. Some go so far as to seek *de jure* veto power for states, which is a recommendation that many thought was implied by IRG. The passage of time has convinced most that a more realistic option is for a potential host state to be given the right to review the federal programs under a negotiated agreement with the agencies and, if necessary, to have objections by the state at a few key decision points considered formally by the President and the Congress. This proposed "consultation and concurrence" power has been advertised as a means of giving the states enough confidence, both in the process and in consistent treatment from the federal government, to remove some of the intergovernmental roadblocks and allow timely decisions to be made on the repository development program. The IRG purists also strongly believe that existing safety and environmental regulatory authorities should be preserved and, moreover, that licensing by the NRC should be required for all potential repositories, including WIPP. In other words, the institutional safe-guards should be applied uniformly to all radioactive waste of equivalent hazard, regardless of whether it comes from commercial or defense programs. In addition, they want a technically conservative approach that includes the investigation of a large number of geologic sites before the first permanent repository is chosen.

This point of view can perhaps be best summarized by quoting an editorial from the *Salt Lake City Tribune,* which addressed the issue of a bill in their legislature that was seeking to give the state veto power over a repository. The editorial recognized that the bill was probably unconstitutional but stated that its real purpose was to get the federal government's attention. It concluded,

> Neither Utah nor any other state can properly refuse to bear the nuclear waste burden once it has been established to the best of human ability that the state offers the most nearly perfect storage conditions. However, the honor of making such sacrifice for time without end must confer on the luckless lamb the satisfaction of knowing first-hand that the duty couldn't have been just as well assigned elsewhere. If Utah's veto stew accomplishes that much it will not have been cooked up in vain.[17]

The IRG purists were greatly disappointed by President Carter's delay in acting on the IRG recommendations and are now afraid that the IRG consensus may have dissipated. President Carter eventually endorsed most of the IRG framework in his policy statement of February 12, 1980; the House and Senate both passed bills incorporating some of its provisions. However, the pressure of other priorities and the political divergence of an election year ultimately prevented Congress from bringing these fragile consensual elements into law.

The Washington pragmatists, which may include some long-term employees of the Department of Energy, take what they consider to be a technocratic approach refined by practical politics. For interim storage, they prefer to increase the federal role in the back end of the nuclear fuel cycle by having DOE take responsibility for the spent fuel filling up utility storage pools. For repository development, they claim that DOE has already implemented the IRG recommendations for an expanded program of research and investigation into several geologic media (besides salt) and for a systems approach incorporating multiple barriers for containment (utilizing the waste form, packaging and geology). To them, steady and visible progress in answering the technical problems will contribute substantially to solving the political problems. They do not believe that comprehensive legislation is needed because a *de facto* political veto power already exists for states and is sufficient to guarantee adequate intergovernmental consultation. Besides, the pragmatists concluded early on that state governors do not really want to take the political heat of a *de jure* veto power. These pragmatists would have also preferred commercial spent fuel rods stored along with military waste at WIPP, but the restrictions forced by the Armed Services Committees—no commercial waste, no concurrence power for New Mexico, and no NRC licensing—can easily be accommodated in order to expedite building a test facility in bedded salt.

Some friends of nuclear power strongly believe that the waste problem would solve itself if a national commitment to nuclear power developed, since the public would then accept the necessity of getting on with the job of permanent disposal. In other words, the waste issue is not a serious technical problem, for enough information now exists to proceed with a conservatively designed repository. Moving ahead with WIPP as a military facility could expedite the first geologic repository and thereby ease the way for other sites. Moreover, segregating military and civilian wastes might help separate nuclear power from bombs in the public consciousness. Although they hoped that federal AFRs would come into being, and that the government would buy and operate the mothballed Barnwell reprocessing plant, they suspected that utilities would probably have to fend for themselves with interim spent fuel storage. The Tennessee Valley Authority (TVA) reached this conclusion early, and any remaining optimists are being forced to reconsider their position in view of the policy pronouncements of the Reagan administration.

The friends of nuclear power are still strongly wedded to the concepts of reprocessing and breeder reactors, but after the unpleasant history of the 1970s are unwilling to accept the financial risk and regulatory uncertainty of commercial reprocessing ventures.[18] The government, in their opinion, has made such a mess of waste management that it might be better to start over and try a federally-chartered or private corporation, like TVA or the Communications Satellite Corporation; indeed, this may be a possible course if DOE is dismantled. The importance of a resolution of the waste problem to the friends of nuclear power was illustrated by their intense lobbying at the end of the 96th Congress for specific measures that involved some flexibility in their own positions. For example, they strongly advocated the legislative sanction of a permanent repository program which included a prescription for a consultation and concurrence process. They temporarily dropped their requirement for federal AFRs in order to get a bill that states and environmentalists would support.

In contrast to the IRG purists and the pragmatists, both the first and fourth groups favor some sort of linkage of nuclear waste management to the future of nuclear power. Ironically, they share a fondness for statutorily defined schedules for the waste management program, one to demonstrate failure and the other to force progress.

These four stereotypical positions do not cover all the views being put forth by important political actors. There has always been a proliferation of small Congressional bills and approaches which do not coincide with any of the above stereotypes and often share only the common theme that Congress be given the ultimate override authority on waste siting. Some bills attempt to institutionalize the IRG and the development of national plans, others to shut down reactors if timetables for waste disposal are not met, and still others to legislate an elaborate framework for consultation and concurrence. Although the IRG process was a remarkable achievement compared with previous efforts to generate societal agreement, its major fault may have been its failure to incorporate key congressional actors early in its deliberations.

The most important congressional deviations can be seen in the actions of the Armed Services Committees in both houses and in the bills perennially proposed by the House Science and Technology Committee and the Senate Energy and Natural Resources Committee. The Armed Services Committees forced the restrictions on WIPP and the separation of the military and civilian repository development programs in DOE. The effort to get a comprehensive bill from the 96th Congress utlimately failed because of the disagreement over a state consultation and concurrence role in defense repositories. The House Science Committee has continued to seek early development of unlicensed demonstration repositories in order to have the symbol of placing large amounts of waste in the ground in this decade, an approach that seems to be gaining favor in the Reagan Administration.

The Senate Energy Committee approach, championed most fervently by Senator Bennett Johnston and other members of the Louisiana delegation, favors giving priority to development of long-term interim storage in engineered facilities above ground rather than to permanent disposal in mined geologic repositories. These retrievable surface storage facilities (RSSF) would be designed for the monitored storage of wastes in a retrievable manner for periods up to or longer than one hundred years. The proponents argue that this technology already exists and will free nuclear power from linkage to finding a permanent solution to waste disposal. It is claimed that the RSSF will also allow flexibility for awaiting new technological developments and eliminate having to make decisions now about the ability of a mined repository to isolate wastes for thousands of years. Although the Atomic Energy Commission temporarily favored this approach after the problems at Lyons, Kansas, all of the efforts of the last four years—including the IRG process—have been directed towards the development of a permanent disposal capability.

Louisiana is the only potential host state with veto power over a repository—a promise given by DOE in order to use Gulf coast salt domes for a strategic petroleum reserve and reaffirmed by both candidates during the 1980 presidential campaign. A cynic might claim that, since Congress can override these presidential promises, the Louisiana delegation's affinity for the RSSF proposal is to ensure that nuclear wastes are never stored in their salt domes. The resurfacing of the RSSF idea, if it gains congressional popularity, may foreshadow continued instability of purpose in the federal waste management program. It also reflects tension that exists between those states with large volumes of temporarily stored waste, such as South Carolina and Idaho that seek transfer of this waste of a permanent repository, and those states that are leading candidates for a permanent repository, such as Louisiana and New Mexico.

Although these caricatures of positions do not convey the richness of views expressed at the Aspen Institute conference, they do serve as a convenient introduction of the diverse competing approaches to how our governance institutions should respond to the nuclear waste disposal problem.

OVERVIEW OF THE CHAPTERS

The foreword by Governor Richard Riley of South Carolina has already ably introduced the central theme of this book and provides a candid statement of the key governance issues from a state perspective. South Carolina has been heavily involved in the nuclear enterprise as evidenced by: five reactors now operating and four more under construction, a commercial low-level waste disposal site that accepted 80 percent of the U.S. total in 1979, the Savannah River Plant that stores large volumes of military high-level wastes, and the partially completed Barnwell commercial reprocessing plant that is a candidate for an away-from-reactor interim storage facility for utility spent fuel. Both as an elected official and Chairman of the State Planning Council on Radioactive Waste Management, Governor Riley has been at the forefront of those state leaders seeking a workable and equitable solution to the radioactive waste management problem. The progress that has been achieved on forging a low-level waste policy consensus (that states should be responsible on a regional basis for ensuring its safe disposal) has been due largely to his efforts in league with the governors of Washington and Nevada. Creating a workable framework for building a federal, state, and tribal partnership in all areas of radioactive waste management has been enhanced by his chairmanship of the State Planning Council and exemplified in the executive summary of its interim report reproduced in Appendix C.[19]

The first chapter by Professor Ted Greenwood of the Massachusetts Institute of Technology looks in depth at radioactive waste management in the United States, with special emphasis on the events of the Carter Administration as well as on the issues with which the Reagan administration must deal. Professor Greenwood brings the combined perspective of a scholar in science policy and a key staff participant in the deliberations of the Interagency Review Group. The comprehensive policy statement that came from President Carter late in his term (reproduced in Appendix A) is founded largely on the ground-breaking work of the IRG. This chapter skillfully explains the intricacies and significance of the governance aspects of nuclear waste management for the country that first embraced the nuclear age.

By comparing waste management policies and programs among the industrialized countries, the second chapter by Dr. David Deese of Harvard University identifies six generic problems: technological bias in decision making; lack of national strategies; fragmentation of governmental power structures; crippled national regulatory bodies; competition among local, state and federal levels of government; and increased importance of nongovernmental actors. In order to strengthen consensus around national programs, he proposes a process that only weakly links progress in waste disposal to the future of nuclear power. The regulatory agencies, in his opinion, have a particularly heavy burden and a key role in developing the technical criteria that assure the safety of permanent disposal and in building a review process that addresses the concerns of various levels of government and the public.

Chapter 3 by Dr. Tom Moss of the House Science and Technology Committee staff and the American Physical Society Waste Management Study critically examines the factional controversies in the last administration and Congress over nuclear waste issues. He concludes that the president was presented by his staff and agencies with political and philosophical choices which did not have to be made at that time, and the resulting delay in presidential action tended to dissipate the IRG's broad consensus on technical issues and to undermine prospects for building coalitions in Congress. At the same time, jurisdictional fixations and hidden agendas prevented Congress from playing a coherent role in formulating a national policy.

The fourth chapter by Mr. Harold Green and Mr. Marc Zell examines the complex legal questions involved in the federal-state conflicts over nuclear waste management. They conclude that attempts by state governments to fill the breach, created by years of regulatory neglect and incompetence at the federal level, by siting moratoria and embargos are likely to be found unconstitutional given the existing legal regime. Nevertheless, they believe that a meaningful role for state participation in the development and regulation of radioactive waste repositories is an idea whose time has come and should be addressed by Congress.

Commissioner Emilio Varanini of the California Energy Commission, in Chapter 5, examines the concept of consultation and concurrence from the perspectives of a host state that is a candidate for a repository and an interested state that has special concerns regarding the demonstration of nuclear waste disposal technology. Using the case of WIPP, he finds that resolution of institutional and technical concerns of New Mexico officials are not being easily accommodated in the existing consultation process due to the lack of state concurrence rights and NRC licensing. He proposes a rigorous process for scientific verification of repository safety that would accommodate the concerns of a state like California which has attempted by statute to link reactor licensing to the demonstration of waste disposal. The California law was upheld in a 1981 federal appeals court decision.

Chapter 6 by Dr. Dorothy Zinberg of Harvard University's Center for Science and International Affairs systematically examines U.S. and European perspectives concerning public participation in nuclear waste management. She finds that the experts will have to listen more carefully to the questions being raised by an increasingly sophisticated public, that has learned to take its rights and responsibilities seriously as participatory democracy is increasingly exercised. In her opinion, it is also becoming important as a part of public education to air the differences within the scientific and science policy community and to clarify the possible compromises that might be negotiated between centrist groups.

Dr. Marvin Resnikoff of the Sierra Club makes a candid and forceful statement in chapter 7 of the goals and tactics of a grassroots environmental organizing effort. From his assessment of past and present practices, he explains why the public should distrust the "politically contaminated science" of the federal agencies. For him, protecting public health and safety will not be achieved through one-sided consensus building efforts, but rather through the democratic process carrying out the public will.

The policy statement of the American Nuclear Society (ANS) on high level radioactive waste disposal given in Appendix B, should be read as an integral chapter of the book. It presents a concise summary of the reasons why this organization of nuclear power proponents concludes that disposal of nuclear wastes in a mined geologic repository is a safe and acceptable approach. The ANS believes that expeditious action is required by the federal government to place into operation a repository for permanent high level waste disposal at the earliest possible time. The ANS is also concerned that delays may impair the authorization of new nuclear reactors.

The epilogue attempts to assess the prospects for consensus in the United States on national policies for radioactive waste management. It is a personal statement rising out of close observation of these issues over a two-year period, but does not represent a consensus by the authors of this book or of the participants at the Aspen Institute conference. All of the chapters in the book should be interpreted as personal assessments.

We would like to thank the Aspen Institute for Humanistic Studies, the Center for Science and International Affairs, and the Kennedy School of Government for their support of the conference and the editing of this book. We are grateful to the authors, whose biographies are in Appendix E, for their clear exposition of nuclear waste politics and to the participants of the Aspen Institute conference, listed in Appendix D, for teaching us about these fascinating governance issues. We would also like to thank Patricia Flaherty for her excellent rapporteur's notes, Mary Ann Spano and Nancy Horton for their superb typing, and David Kellogg and Jack Hoffman of Pergamon Press for their adroit shepherding of this manuscript. We wish to acknowledge as well our benefit from having observed the skillful efforts aimed at building societal consensus on waste disposal issues by the Keystone Center and the Conservation Foundations RESOLVE Forum. Lastly, we would like to express our deepest gratitude to the Mitre Corporation and the Electric Power Research Institute, both for their generous and "no strings" financial support of the conference and their patience in awaiting this book.

NOTES

1. Interagency Review Group on Nuclear Waste Management, *Report to the President by the Interagency Review Group on Nuclear Waste Management,* March 1979, TID-29442, p. 87. For interesting reviews of the history of waste management, see Richard G. Hewlett," Federal Policy for the Disposal of Highly Radioactive Wastes from Commercial Nuclear Power: An Historical Analysis," U.S. Department of Energy, Washington, DC, March 9, 1978, and William W. Hambleton, "Historical Perspectives on Radioactive Waste Management" (by the Director of the Kansas Geological Survey), presented at a symposium in Madison, Wisconsin, April 24, 1981.

2. U.S. Department of Energy, Final Environmental Impact Statement on the Management of Commercially Generated Radioactive Waste, October 1980, DOE/EIS-0046F.

3. Hambleton, "Historical Perspectives."

4. Hewlett, "Federal Policy for the Disposal of Highly Radioactive Wastes." This paper shows that the fundamental questions and issues of waste management have endured for decades.

5. See Appendix A of this volume.

6. Public Law 96-164, December 29, 1979.

7. Albuquerque Journal, January 24, 1981.

8. Albuquerque Journal, January 26, 1981.

9. Stipulated Agreement, U.S. District Court for the District of New Mexico, Civil Action 81-0363 JB, July 1, 1981.

10. U.S. Department of Energy, "DOE Announces New Spent Fuel Nuclear Policy", press release, October 18, 1977.

11. U.S. Department of Energy, *Spent Fuel Storage Requirements—The Need for Away-From-Reactor Storage,* DOE/NE-0002, January, 1980. The most recent official projections, which are lower, are given in the letter from Omer Brown, Attorney, Office of the General Counsel, Department of Energy, to Marshall E. Miller, Administrative Judge, U.S. Nuclear Regulatory Commission, March 27, 1981.

12. Richard Riley, speech in April 1980 at Harvard University.

13. See letter from Omer Brown in reference 11 and an unpublished, but widely circulated memorandum from the President to the Secretary of Energy, March 20, 1981.

14. The Low Level Radioactive Waste Policy Act, Public Law 96-573.

15. The Congress authorized cleanup of certain inactive mill tailing sites in the Uranium Mill Tailings Radiation Control Act in 1978.

16. Interagency Review Group, *Report to the President.*

17. Salt Lake City Tribune, April 24, 1981.

18. Atomic Industrial Forum, "A Policy Paper on AFR's and Reprocessing", May 18, 1981, and unpublished minutes of the meeting on reprocessing between the acting Undersecretary of Energy and representatives of the nuclear industry, April 23, 1981.

19. The final recommendations of the State Planning Council on Radioactive Waste Management are included in its *Report to the President,* August 1981.

1

Nuclear Waste Management in the United States

Ted Greenwood

Nuclear wastes comprise one category within a general class of hazardous wastes that modern industrial activity forces society to manage,[1] and in some respects they are neither unique nor particularly noteworthy within that class. They are neither the only long-lived hazardous wastes nor the only ones whose level of toxicity requires management by containment and isolation rather than by dispersion and dilution. Noting this fact does not imply that managing nuclear wastes either is or should be simple or straightforward. The management of all hazardous wastes is difficult, and the historic record in the United States is by no means a source of national pride. Moreover, nuclear wastes do have certain unique characteristics that render their management even more difficult than usual.

First, their hazard derives primarily from their radioactivity, an invisible, elusive, and not easily understood property of matter that is beyond the direct experience of most people. Second, in the case of high-level nuclear waste—the category that receives the most attention—public expectations and regulatory standards require high confidence that isolation can be sustained for thousands, if not tens of thousands of years. This implies the need for a level of understanding and capability for predicting the behavior of complex geological systems that skeptics can readily argue does not exist. Third, most of the existing and future nuclear waste is and will be the by-product of activities with which much of the public associates the risk of disaster, namely, nuclear weapons production and nuclear power generation. Both are extremely controversial activities against which extensive political opposition has been mobilized. As a result of these special characteristics, nuclear wastes evoke in the minds of many an aura of mystery, an association with destruction and human

1

suffering, and a hesitancy to accept assurances that disposal can be carried out safely.

As with most other societal activities that involve safety, health, or environmental risks, several technical approaches are available for disposal of the various categories of nuclear waste. These differ in their levels of cost, their degrees of associated risk, and how and to whom those costs and risks are allocated. Rarely can the costs and risks of any particular action be determined precisely. Controversy exists on a generic level about the various methodologies used for cost and risk assessment and on a case-specific level about whether any particular assessment has been done correctly and completely. Rarely can total risk be reduced without increasing cost. No general agreement exists about how or even whether costs should be traded against risk reduction. Sometimes, near-term costs can be avoided with little or no increases in near-term risk by transferring risk to future and perhaps distant generations. Whether such intergenerational risk trade-offs are acceptable or should be analyzed within an economic paradigm using standard discounting procedures is controversial. The analysis of risks, costs, and benefits requires information, much of which is highly technical. But information may be unavailable for use by policy analysts or decision makers for several reasons: it may not yet be known; it may be known but not accessible when needed; or it may be unknowable even in principle. In any event, some of the information used in risk/benefit and cost-benefit analysis is likely not to be universally accepted as correct or even pertinent.

To the extent that governments become involved in regulation of private sector management of hazardous materials, as they have increasingly during recent decades in the United States, additional complexities arise. As always, the institutional and political setting in which regulatory decisions are made has important implications for decision outcomes. A government regulatory body's internal structure and procedures, the modes of interaction between government regulators and interest groups and among various interested government entities, and the role of the scientific and technical community and the general public, all make a difference. Participants in regulatory matters understand this to be true, and as a result, argue and lobby vigorously over questions of process, procedure, and structure as well as over questions of data and technical judgment. Indeed, the former sometimes so dominate an issue that participants appear to believe that technical accuracy is either unimportant or asssured, given particular configurations of process, procedure, and structure. These factors apply strongly to the nuclear waste issue because both nuclear waste management—including storage, transportation, and disposal—and the commercial activities that generate such wastes come under the purview of not just one but several regulatory agencies in the United States. Indeed, the commercial nuclear industry is and always has been perhaps the most highly regulated industry.

In the United States, as elsewhere, government is involved in nuclear waste management, not only as regulator; the federal government also manages and will conduct the disposal of large quantities of the waste. It is responsible for both interim storage and disposal of all wastes generated in government national security and research and development programs. Under existing law, it also is responsible for the disposal of all high-level wastes generated in the private sector. In addition, disposal of large quantitites of waste products from discontinued private uranium mining and milling operations must be carried out by the federal government.

As a result of the government's operational role, further complications are introduced. Within the restrictions of personnel ceiling and civil service constraints, the government must create and maintain an operating organization with appropriate managerial and technical skills. Funds must be allocated by means of the highly competitive and politicized budgetary process. Program activities and decisions about the allocation of costs and risks are subject to political influence, patronage, the whims of bureaucratic managers or legislative overseers, and the need to take into account the wishes of powerful interest groups. The authorities of various segments of the bureaucracy are overlapping and sometimes in conflict. Each agency has its own organizational interest and political constituency. No single-value measure of programmatic success exists, such as profitability in the private sector. Shifts in national mood or priorities frequently require altered emphasis and sometimes total program reorientation. Account must always be taken of the possible impacts of actions in one area on policies and programs in other areas, including relations with other countries.

In addition to these complications which apply to any government operating program, the unique features of the nuclear area further exacerbate the government's problems. The strong public anxiety about and concern with nuclear power, the attention given to it by the press, opinion leaders, and environmental activists, and the resultant political mobilization around the issue inevitably influences discussion of the nuclear waste issue. Not only do those advocating reduced or no reliance on nuclear power see the waste as a serious hazard that the world would be better off without, but they also exploit the waste issue as an effective instrument in their struggle against nuclear power operations. Similarly, those opposed to nuclear weapons production find that focusing on the hazards of the resultant nuclear wastes is an effective, although indirect, means of exerting pressure on the weapons-related activities themselves.

The U.S. nuclear waste management program is at a critical decision point in 1981. Over a period of two years beginning in 1978, the Carter administration conducted a detailed and comprehensive reevaluation of this complex and troublesome issue. The result was a call for significant policy reorientation and associated redirection of program activities, as outlined in a presidential message to Congress on February 12, 1980 (This message, reproduced in Appendix

A, will be referred to as "President Carter's Nuclear Waste Message"). Some of the proposed changes were implemented before President Carter left office, but not all. Some were blocked by Congress and others by executive agency opposition or inactivity. The Reagan administration and the Congress are looking anew at the nuclear waste issues. At this writing the near-term orientation of the national effort has still not been fully defined. The major question both the administration and the Congress must address is whether to proceed along a course similar to that set by President Carter or to strike out yet again in another direction.

The major purpose of this chapter is to pose and analyze the major choices that must be faced during the term of the new administration. To be able to do this, however, attention must first focus briefly on the historical background of the radioactive waste issue in the United States, the organizational and institutional context within which the waste management programs exist, and the underlying assumptions and philosophy of the Carter administration's approach to the issue. To address the latter subject, the process and outcomes of the Carter administration's policy review will be discussed.

HISTORICAL BACKGROUND[2]

Long before the launching of the commercial nuclear industry in the United States, large quantities of radioactive waste were generated by the national security activities of the Atomic Energy Commission (AEC). Responsibility for the management of this waste also resided within the AEC. However, because the agency was oriented toward the development and production of nuclear weapons and was supported in this perspective by the Congress and the national mood, waste management initially received relatively little attention at the policy level. Large quantitites of waste were disposed of and interim storage choices were made for the rest, including, for example, the placement and later neutralization of high-level wastes in corrosion-prone carbon steel tanks. However, some of these measures were taken in absence of a firm technical basis and with what now seems to have been insufficient or incorrect forethought.

Over time, changes were made in techniques used for processing and storage of high-level and transuranic wastes, driven primarily by the related objectives of reducing volume and cost. For many years there seemed to be little urgency to move toward permanent disposal, and the belief was widely held within the AEC that safe disposal would be possible at the three AEC reservations at which the waste was generated and stored in quantity: Idaho Falls, Idaho; Hanford, Washington; and Savannah River, South Carolina.

Extensive private uranium mining and milling operations gradually developed in the 1940s and 1950s to supply the AEC national security activities.

Government oversight of this industry's waste management practices was minimal in the early years, however, with the result that companies used low cost but not always the safest methods. As a result, massive and expensive remedial action must now be taken at older mill sites, both operating and abandoned. Gradually, regulation of the uranium industry was strengthened, but relatively little attention was given by either government or industry to generating the basic information necessary to evaluate the safety of various approaches for managing tailing piles or to investigating and developing improved techniques.

Low-level wastes from AEC activities were disposed of by shallow burial on AEC land without independent regulatory oversight. Commercially generated low-level wastes were buried at similar, commercially run sites under the regulatory control of the AEC and later the Nuclear Regulatory Commission (NRC) or of state authorities acting under agreement with the federal agency. In this area, too, little research and development (R&D) was conducted on new methods of safe disposal. In retrospect, some of the existing sites and methods have turned out to have been chosen with less care and consideration of long-term effects than should have been given or that probably would be given today. The financial arrangements for guaranteeing long-term care of some of the commercial sites have also turned out to have been inadequate.

With the enactment of the Atomic Energy Act of 1954, the legal basis for launching a commercial nuclear industry in the United States, the need emerged to think about the disposal of the high-level radioactive wastes this industry would produce. Because legal requirements existed for regulatory overview of commercial activities which did not apply to government activities, storing commercial waste together with government waste was precluded. Within a few years the AEC had developed and adopted a management and disposal concept for planning purposes. It assumed that commercial spent fuel would be stored only briefly at reactors where it was generated and then would be reprocessed; that is, usable uranium and plutonium would be chemically extracted from the spent fuel. The high-level waste product of reprocessing operations would be stored for some time at the reprocessing plant, then solidified, appropriately packaged, and eventually disposed of in an underground cavity (called a repository) mined out of rock salt. Disposal in salt deposits had been proposed in 1957 by a committee of the National Academy of Sciences as the "most promising method of disposal for high-level" waste at the present time.[3] This reference disposal concept persisted with little change into the 1970s and was fundamentally identical to the concepts for disposal of high-level nuclear waste developed in Europe.

In its efforts to launch a commercial nuclear power industry, the AEC was concerned primarily with the power reactors themselves—light water reactors first, liquid metal fast breeders later—and only secondarily with the back end of the fuel cycle. The exciting, challenging, and rewarding work was associated with reactors, and these commanded most of the time and attention of the agency's best technical people and its senior management. The AEC did take actions to encourage the construction of commercial reprocessing plants, but until the mid- to late 1960s did relatively little to put in place a disposal facility for high-level waste. This is not particularly surprising. The technical problems of disposal were not considered to be very great and, under the AEC's baseline concept, no commercial waste would be available for disposal until reprocessing operations were underway. Moreover, no real urgency was perceived to dispose of the existing defense waste. Indeed, delay would permit the waste's thermal output to decline as a result of natural decay processes and would make subsequent disposal in a repository easier and more efficient.

Organizationally, the commercial nuclear waste program was quite separate from the defense-related nuclear waste activities within the AEC. This organizational structure had the disadvantage of inhibiting integrated and consistent planning, although some common contractors were employed and some technical information was exchanged. In the public mind, however, the two programs were closely linked. Leaks at the storage tanks at Hanford and migration of radionuclides at government low-level waste burial sites, although they resulted in little or no actual harm, seriously damaged the credibility of the AEC and its successor agencies as trustworthy guardians of the public health and safety. The newly created Energy Research and Development Administration (ERDA) organizationally joined the two waste management programs in 1975 and placed the new operation within the commercial nuclear R&D segment of the agency. However, their congressional oversight—and therefore their budgets—remained divided. Like other program sectors of the technically oriented ERDA, the commercial nuclear R&D effort was primarily interested in developing and promoting the energy-producing technologies under its purview. Nuclear waste activities continued to have lower priority.

In both the AEC and ERDA, attention was focused on technical aspects of the nuclear waste problem to the virtual exclusion of institutional and political aspects. This was most striking in the high-level waste area, where the problem became defined as locating an appropriate site for construction of one salt repository. A much-touted plan to construct the first repository in a salt formation near Lyons, Kansas was abandoned in 1971 because of technical inadequacies and political opposition. Because the AEC had taken inadequate care in characterizing the site prior to making programmatic commitments to it and had insufficiently consulted with the state's political authorities, the state Geological Survey, and the general population, this failure reflected very unfavorably on the agency's competence and trustworthiness.

Following the Lyons repository failure, the AEC redirected its efforts toward long-term surface storage of high-level waste. A draft environmental impact statement on this proposal, issued in September 1974,[4] drew heavy criticism, however. Environmental groups, state and local governments, and the Environmental Protection Agency argued that the proposal seemed to place in abeyance efforts directed toward permanent disposal and that a significant possibility existed that the storage, represented as long term but temporary, would by default become permanent. The following April, ERDA withdrew the document and abandoned the concept of long-term surface storage in favor of proceeding immediately with geologic disposal.

The waste management concept and programs of the AEC and, to a lesser extent, of ERDA, were narrowly conceived and lacked both flexibility and redundancy. When the proposal to build a repository at Lyons collapsed, the AEC had no alternative sites and no fallback position. The entire high-level waste disposal effort came to a halt while the commission struggled to find a new approach. The resultant proposal for retrievable surface storage also stood alone, and when it was subsequently withdrawn by ERDA, the planning had again to start afresh.

In late 1976 ERDA did propose a diversified geologic strategy that included a broad search in thirty-six states for acceptable repository sites and ultimately for the construction of several facilities.[5] However, it met growing hostility on the part of state governments to the idea of high-level waste disposal facilities or even to geologic investigations within their borders that might conceivably locate a technically suitable site. Several states even tried officially to prohibit activities related to waste disposal. These actions raised starkly the issue of states' rights versus federal preemption in the area of nuclear waste disposal, an issue still unresolved today. In the face of such political opposition and impediments to progress at the state level, ERDA soon drew back into a narrower approach of searching for a single salt site in a state that would not object to a repository within its borders.

Within the technical components of the high-level waste disposal program, AEC and ERDA efforts were focused strongly on the engineering aspects of the problem, particularly the design and construction of the repository and the development of an adequate waste form. Risk assessment methodology and important earth sciences issues, particularly in the fields of geochemistry and hydrogeology, were given inadequate attention. However, the scientific basis of the waste disposal programs came increasingly under scrutiny by environmental groups, the U.S. Geological Survey, the Environmental Protection Agency (EPA), and the NRC. Concerns were raised and gradually spread that existing knowledge was not sufficient to provide high confidence of long-term isolation and that technical assumptions on which the reference disposal concept was based (for example, the durability of borosilicate glass under repository conditions in a salt medium, the absence of water in rock salt, and the suitability of all DOE reservations for disposal of defense wastes) may not be correct.

Each of these occurrences was a serious blow to the morale and public credibility of the nuclear waste disposal program. They resulted in large measure from the government's failure to mount a technical program with sufficient diversity and scientific underpinnings and to deal adequately with an ever-more complex political and institutional environment. As this environment became so hostile that progress was seriously impeded, the program's managers gradually did turn their attention to protecting and repairing relations with state and local governments, to taking account of the views of special interest groups and interested members of the public, and to expanding relations with other federal agencies that possessed relevant expertise and authorities. However, the responsible government officials were unable to understand sufficiently what was happening and to take appropriate action to rebuild public confidence. The result was public skepticism, if not hostility, toward the government's program, and growing paralysis of it. Each of these fed on and reinforced the other.

Government agencies and officials should perhaps not be greatly faulted for having adopted a narrow view of the radioactive waste disposal problem, particularly in the early days. This was in keeping with the perspective of the times and the relative priorities of the public agenda in the 1950s and most of the 1960s. These were the days of technological optimism before the environmental awakening of the late 1960s, the days when nuclear power was almost universally perceived to hold the exciting promise of abundant, cheap, and safe electric power.

But national priorities, perspectives, and agendas gradually changed. The public's trust in technology as the unblemished source of prosperity and beneficence and in the government as the faithful guardian of the public's welfare eroded. The National Environmental Policy Act (NEPA) became law, imposing strict procedural requirements on all government programs and providing ready mechanisms for nongovernment critics to challenge programs and proposals. The AEC gave way to ERDA and the NRC, the latter being established explicitly to be an independent watchdog on the nuclear promoters in government and industry. EPA was given authority to promulgate general guidance and standards for radiation exposure to be used in regulation by other governmental agencies. The Department of Transportation (DOT) became a regulatory agency with oversight over the movement of hazardous materials. However DOT failed to exercise its authority to regulate the transport of nuclear wastes on the nation's roads. ERDA in turn gave way to the Department of Energy (DOE), and the immediate responsibility for the nuclear waste program was elevated to a more visible and authoritative position at the deputy-assistant-secretary level.

During this period, nuclear power passed through its initial honeymoon period and became the subject of a major national debate. Powerful, organized, and sometimes well-financed opponents began employing all available instruments of political action—judicial challenges, demonstrations, legislative proposals, state referendum initiatives, financing contenders for public office—to muster support for their cause of slowing down new nuclear plant construction or even shutting down operating facilities. The rallying causes of the nuclear opponents have been nuclear reactor safety, the potential for nuclear weapon proliferation that many feared could result from the widespread use of plutonium, and concerns over nuclear waste disposal. Until the accident of Three Mile Island redirected national attention toward nuclear plant safety, nuclear waste had emerged as the primary—and a quite effective—focus of the nuclear power opponents. Although somewhat belatedly, opponents of nuclear weapons production also began to use the waste issue in an effort to close down or impede the government's nuclear national security enterprise.

As the external environment changed, the waste disposal programs did not adjust adequately. Some useful efforts were made during the Ford administration, including allocating greater resources to the high-level waste disposal problem and expanding the site characterization program. But their effectiveness was constrained by organizational impediments, by a history that almost yearly became a longer litany of failures to be redressed, explained away, or lived down, and by the absence of management personnel able to overcome these difficulties. In addition, repeated internal reorganizations and restructuring of parent agencies had debilitating effects on programs and personnel.

When the Carter administration took office in 1977, the high-level waste disposal program was in serious trouble. Technical progress was still being made, but, in political terms, the government's program managers seemed to be steadily losing ground against their critics. Nuclear power appeared in jeopardy because of a widespread perception that the government did not know how—or at least was not doing much—to solve the waste disposal problem. Several states had acted legislatively, though specific arrangements with ERDA, or otherwise to try to remove themselves from consideration as potential hosts for a high-level waste repository, to prevent nuclear wastes generated elsewhere from crossing their borders, or to link the question of nuclear power plant licensing to waste disposal. Local governments, too, had begun to enact laws and ordinances that regulated and sometimes impeded the highway transport of nuclear wastes.

In July 1977 the NRC denied a petition of the Natural Resources Defense Council requesting a rule-making proceeding to determine whether high-level waste can be permanently stored without undue risk to the public health and safety and a moratorium on reactor licensing until an affirmative determination had been made. In doing so, the commission stated that it had "reasonable assurance that methods of safe permanent disposal of high level wastes can be available when they are needed." However, it also indicated that "the Commission would not continue to license reactors if it did not have reasonable confidence that the wastes can and will in due course be disposed of safely."[6] By including as conditions of continued reactor licensing not only a requirement for technical feasibility but also a requirement that progress be made in the government program, the commission put additional pressure on the program managers.

The back end of the commercial nuclear fuel cycle was in disarray by late 1976 because of economic problems associated with reprocessing, the technical failure of one industrial attempt to operate a reprocessing plant, and both changes and continuing uncertainties in the regulatory environment for reprocessing and the commercial use of plutonium. The single reprocessing plant in the United States, that had operated as a commercial venture, near West Valley, New York, closed down for reconstruction and expansion in 1972, but the owners later decided not to reopen it. Over 600,000 gallons of high-level waste in storage tanks at that facility await processing and disposal. Moreover, there was a growing perception in the United States that reprocessing was not economically attractive and should be delayed or foregone in order to avoid the risks—mostly related to nuclear weapons proliferation—involved in the use and transport of plutonium.

The Carter administration recognized that the nuclear waste issue needed careful policy-level review, and acted within its first year to initiate such review. Before proceeding to a description of the process by which this review was conducted and an analysis of its results, however, additional background on the existing organizational and institutional context of the U.S. nuclear waste management program must be presented.

THE CURRENT ORGANIZATIONAL AND INSTITUTIONAL CONTEXT

The existing organizational and institutional structure of the waste management program in the United States is quite complex; it involves not only a variety of government agencies, but also the Congress, the courts, state and local governments, and a host of interested and relevant nongovernment entities. This section will provide a brief explanation of the responsibilities, functions, and points of leverage of each of these organizations.[7] These are significant because, as stressed earlier, government structure, the allocation of responsibilities, the processes through which issues get addressed, and the iinkages with interested groups and specialists outside the government are important determinants of policy outcomes, and they strongly influence day-to-day program activities.

The federal government's nonregulatory responsibilities for radioactive waste management and disposal reside with the Secretary of Energy.[8] Program responsibility resides in the Office of Nuclear Waste Management under the Assistant Secretary for Nuclear Energy. This office is authorized for ninety-eight full-time personnel in FY 1981, most of whom are headquartered in Germantown, Maryland. The rest are at field offices around the country. Its primary functions are planning, budgeting, coordination, integration, and program analysis. Virtually all programmatic activities are conducted by contractors who report to various DOE regional operations offices. This decentralized management structure has some advantages in that it mobilizes capable talent all over the country and gives considerable authority and independence in program execution to operating units. However, it makes coordination, quality control, and consistency among various elements of the program somewhat difficult to achieve. In the high-level waste program, for example, comparison among potential repository sites is a necessity. However, despite the existence of a lead contractor—Battelle Memorial Institute in Columbus, Ohio—responsible for most of the geologic investigations and for preparing standardized criteria, procedures, and techniques for site characterization, some sites are being studied by different organizations largely using procedures and techniques they developed themselves. This management structure will make site intercomparison difficult to carry out. DOE recognizes the problem and, for this and other reasons has been increasing the number of technical people in its Columbus, Ohio office.

Federal expenditures for nuclear waste management increased dramatically under the Carter administration. ERDA's expanded program was budgeted at $237 million in FY 1977 and was itself a major increase over FY 1976. DOE's new budget authority for nuclear waste in FY 1979 rose to $483 million. Since then, expenditures have been about level in real terms, with $511 million in FY 1980 and $569 million in FY 1981. The defense components of the nuclear waste budget, driven primarily by requirements for interim care of existing defense wastes, declined steadily in percentage terms from 1971, when it comprised the total appropriation, until 1981, when it represented 53 percent. This relative decline resulted from the initiation of the remedial action program in 1976 and a small spent-fuel program in 1978 and the steady strengthening of the commercial high-level waste program which jumped from 10 to 28 percent in FY 1977 and reached 36 percent in 1981.

The Nuclear Regulatory Commission (NRC) is an independent regulatory agency established in 1975 to oversee commercial nuclear activities in the United States. Under current law, it is responsible for licensing all waste handling, transportation, processing, and disposal activities involving all kinds of commercially generated nuclear waste, whether such activities are actually conducted by the private sector or by government, and the disposal of government-generated high-level wastes. NRC does not have jurisdiction over government activities that generate waste, government waste management research and development activities, or the disposal of government-generated low-level or transuranic wastes. NRC is to set standards that serve as the basis for licensing, to actually carry out licensing, to prepare environmental impact statements to support both standard setting and licensing activities, to work with states in all aspects of waste disposal, and to conduct R&D required to support its licensing actions. Like DOE, most of NRC's programmatic activities are conducted by contractors, including the national laboratories administered by DOE. The size and budget of NRC's waste management activities have grown rapidly in recent years. Full-time government personnel rose from 120 in FY 1980 to 140 in FY 1981, and President Carter's budget called for 152 in FY 1982 for work on high-level waste, low-level waste, and uranium mill tailings. The program support expenditures rose from $17 million in FY 1980 to $23 million in FY 1981 and $40 million in the FY 1982 request. A little more than a third of the personnel are allocated to high-level waste, but more than half the program support expenditures are in that area.

The Environmental Protection Agency (EPA) is a regulatory body with broad standard-setting and enforcement responsibilities. In the nuclear waste areas, its primary role is to set standards for radiation exposures of the general public that must be adhered to by other federal agencies, particularly DOE, NRC, and DOT in their own activities. In particular, NRC must follow EPA guidance and implement EPA standards in its regulation of DOE and commercial waste disposal activities. In recent years, EPA's level of effort in the nuclear waste area has held about level at 30 full-time employees and $2.0 million.

The interlocking of authority and responsibility among EPA, NRC, and DOE should be relatively straightforward: EPA sets general standards; NRC implements these in the establishment of its own detailed criteria and licensing standards; DOE responds to NRC standards and criteria for its activities that NRC licenses and to EPA standards for those activities that are not licensed. These relative dependencies should also determine the order in which actions are taken. In reality, however, DOE's activities—disposal of low-level wastes and planning for disposal of high-level and transuranic wastes—are ongoing and much action has been taken in the past, but most of the EPA and NRC standards and criteria relevant to DOE activities do not yet exist in final form at this writing. Moreover, current schedules would indicate that NRC may take action in several areas before EPA, although EPA's standards must be complied with when they are issued.

This situation creates a number of problems. First, DOE and NRC are proceeding in the absence of the standards that they must ultimately meet. Despite regular informal interagency coordination, some risk exists that changes will later need to be made in NRC standards and DOE practices. Moreover, NRC's standard setting task is made much more difficult because of the absence of EPA guidelines. Second, NRC and EPA may be, and certainly are perceived to be, under pressure not to promulgate standards that would interfere with current DOE plans or require expensive remedial actions. Third, in areas where DOE requires an NRC license or must act consistent with EPA standards, its planning and ongoing program activities are held hostage to what NRC might ultimately decide.

The Department of Transportation (DOT), through the Materials Transportation Bureau, has regulatory authority over the transportation of nuclear wastes. To deal with the overlap of authority with NRC, a memorandum of understanding[9] exists between the two agencies. Under it, NRC develops safety standards for design and performance of packages for transport of fissile materials and quantities of radioactive materials exceeding certain limits. DOT then enforces these standards and sets its own qualifications for carrier personnel and handling procedures.

The Department of the Interior (DOI) has two responsibilities in the area of nuclear waste management: providing earth sciences expertise and supplying parcels of federal land that may be selected for study or, ultimately, for facility sitings. The U.S. Geological Survey conducts laboratory R&D and field geologic investigations in support of DOE's high-level waste disposal program and, to a lesser degree, under its own authority in order to supplement DOE's program. DOE pass–through funding began mondestly in FY 1975, and in FY 1981 was $5.5 million. The Survey's own program related to high-level waste began in FY 1979 and was $4.4 million in 1981. The Survey may also be asked by NRC and states to act in an advisory capacity when they evaluate DOE applications for the construction of repositories. In the area of low-level waste, the Survey's own program was at the level of $2.3 million in recent years and $0.6 million was provided by DOE. Personnel for the total nuclear waste effort has risen steadily, reaching 87 full-time and 111 part-time people in FY 1981.

The Bureau of Land Management must be an active party to any proposal to site a waste disposal facility or to open a uranium mine or mill on any portion of the federal government's large land holdings under the bureau's custody. In fact, much favorable geology for high-level waste disposal and many of the uranium deposits in the United States underly public lands. Uranium is also found on Indian land, some of which is under the oversight of the department's Bureau of Indian Affairs.

The Department of State is involved in waste management issues to the extent that they relate to nonproliferation policy, involve cooperative R&D or other activities with foreign countries, or attract the interest of and influence actions taken by other countries. Particularly in recent years, high-level waste storage and disposal, like many other aspects of domestic nuclear policy, have become inextricably bound together with nonproliferation policy. Domestic actions are sometimes taken to foster or support foreign policy objectives, and actions taken or not taken by other countries sometimes have direct or indirect influence on U.S. domestic programs.

The precise roles of the various presidential advisors and staff units within the Executive Office of the President (EOP) vary according to the preferences and style of operations of each president, but several are always particularly relevant. The president's political strategists, his press secretary, the Intergovernmental Affairs Office, and his domestic policy advisors are interested in nuclear waste matters to the extent that they have significant political implications for the president. Particularly the latter, as coordinators of domestic policy issues for the president, can have a strong if not dominant policy role whenever they choose to exert their influence. The Intergovernmental Affairs Office is interested in the critical questions of federal-state relations in the nuclear waste area. The congressional liaison staff also can play an important role because it serves both to predict and to interpret congressional actions within the White House and to orchestrate efforts to obtain congressional implementation of the president's policy.

The Office of Management and Budget (OMB), as the watchdog for and coordinator of the president's annual budget, can exert considerable influence over the scale and orientation of agency programs. In areas where the president's political advisors choose not to be deeply involved, OMB can have the dominant role within the Executive Office. It serves as a coordinator or broker between and among agencies when the usual processes of interagency discussion and bargaining break down. It also coordinates legislative proposals and congressional testimony to ensure consistency and compatibility with the president's policy, although it does so with uneven effectiveness. OMB, of course, has a strong bias to address all issues in budgetary and personnel quota terms and to restrain growth in federal spending.

The Council on Environmental Quality (CEQ) is mandated by law to act as overseer of the National Environmental Policy Act and can be expected to take a special interest in nuclear waste so long as it remains high on the priority list of nongovernmental environmental organizations which form CEQ's constiuency. CEQ exercises power in the nuclear waste area through its roles as consultant to OMB, advisor to the president, channel of information and influence for the external environmental community, and official watchdog of the environmental impact statement process. CEQ's effectiveness, however, depends totally on the extent to which the president looks to it for advice and is seen by others to do so.

The Office of Science and Technology Policy (OSTP), an office that reemerged in 1976 after President Nixon's elimination of its predecessor in 1973, has a special interest in the technical and scientific aspects of the waste management programs. Its influence is exercised through its roles of consultant to OMB, advisor to the president, channel of information from the country's scientific community, and through its ability, like OMB, to act as coordinator and broker among the agencies. OSTP's effectiveness also depends on its relationship with other EOP offices and on the extent to which its director is relied upon by the president.

The National Security Council staff, like the State Department, is interested in waste management to the extent that it impacts nonproliferation policy.

The United States Congress, of course, is a unique legislative body in that it is totally separate from the executive branch and, unlike many national parliaments, can and does act independently of and frequently at cross-purposes to the executive. Any alteration in the statutory framework of the waste management program—including, for example, changes in the breadth of NRC's regulatory oversight, changes in the relative authorities and legal powers of the federal and state or local governments—requires an act of Congress. Legislation addressing bureaucratic structure or general policy directions can be submitted to Congress by the president or initiated by the appropriate congressional committee. In the former case, the Congress rarely adopts presidential proposals without change. The president, of course, can veto bills he sufficiently dislikes, but the political costs may be high, the disruption to the nuclear waste program or other programs covered in the same bill may be extensive, and his veto can be overriden by a two-thirds vote of both houses of Congress.

Congress and particularly its committees also exert great leverage over policy and programs through the annual budget process. After the president sends his budget request to Congress, the latter must first authorize funds to be spent for particular purposes and then appropriate the money. In the process, very substantial change can and frequently is made from the president's proposal. One committee of each house is responsible for the appropriations bills, but four in each house are involved in the authorization of nuclear waste programs and therefore in waste management policy. These committees frequently disagree on policy and programs, constantly squabble over jealously protected authorities and jurisdictions, and regularly serve as channels through which special interest groups, regional or local preferences, or pet projects of the members find expression and influence. Great bureaucratic and political skill is needed on the part of any executive branch program manager to ensure that his programs survive the congressional budget process close to the way the president submitted them or, if he is careful and prudent or simply powerful, considerably better from his own point of view. For an issue as contentious as nuclear waste, the congressional decision-making process provides an opportunity for all perspectives to find expression and involves lively debate among committees and powerful individuals and between Congress and the executive branch.

State governments also have significant roles in nuclear waste management even though, under the Atomic Energy Act, most authority over nuclear matters, including the regulation of high-level waste disposal, has been preempted by the federal government.[10] Indeed the States have in recent years sought both to clarify and to increase their role in planning and decision making for nuclear waste activities. Under legislation enacted in December 1980, states have the responsibility for providing the capability for disposal of low-level nuclear waste. In addition, the Atomic Energy Act permits NRC to enter into agreements with states through which the latter are delegated responsibility to license low-level waste disposal, uranium milling operations, and decommissioned facilities within their boundaries. There are twenty-six such agreement states. Delegation or even sharing of authority by NRC in the case of high-level waste is not possible under current law. The questions of whether states should be given the authority to prevent siting of a nuclear waste facility that the federal government wants to build and, if so, how that authority should be exercised and what its limits would be are subjects of lively current debate. They raise difficult legal and constitutional questions that have been faced in many other areas, rarely without difficulty. Ultimate resolution will probably require an act of Congress. Nonetheless, many people believe that as a practical matter, states almost certainly could prevent or at least make very difficult the siting of a high-level waste repository within their borders if they choose to try.

Waste transportation, particularly by highway, is an area where states and local governments have acted to impose constraints, including total bans. However, on January 19, 1981 the Department of Transportation published a final rule, to take effect February 1, 1982, that would impose a uniform federal regulatory regime nationwide and preempt most state actions on highway transport of nuclear wastes.[11] States would, however, share with the Department of Transportation the authority for designating approved highway routes for the transporting of nuclear wastes. Whether the states' preference that a prenotification requirement be required or permitted under the rule remains unsettled pending further review by DOT and NRC.

Some states have also acted to link the nuclear waste issue to the nuclear power plant siting question by imposing a moratorium on new plants until a waste disposal facility is available or until they have greater confidence that the nuclear waste generated by such plants will in due time be safely disposed of. However, the prototypical California law was struck down in 1979 by a federal district court and is now being appealed.[12]

The courts also play an important role in the waste management programs. Virtually all actions of the executive branch are subject to judicial review and can be challenged in court. In the nuclear waste area, they are challenged regularly. The National Environmental Policy Act is particularly significant in this regard because it imposes certain procedural requirements for decision making and requires that adequate environmental impact statements (EISs) be prepared prior to any major action of the federal government that would have a significant environmental impact. Since virtually all actions related to nuclear waste have or could have significant environmental impact, EISs are produced at virtually every turn. These can be and are likely to be challenged in court on the issue of adequacy.

Licensing and standard-setting activities of regulatory agencies can also be and regularly are challenged in court. One very significant recent court action was that of the United States Court of Appeals for the District of Columbia Circuit in *State of Minnesota* v. *NRC* (May 23, 1979).[13] As a result of the court's statement, NRC decided to conduct a proceeding "to reassess its degree of confidence that radioactive wastes produced by nuclear facilities will be safely disposed of, to determine when any such disposal will be available, and whether such wastes can be safely stored until they are safely disposed of."[14] This so-called confidence proceeding is underway at this writing and will be referred to below. Finally, attempts by states to restrict federal activities or to prevent wastes generated elsewhere from entering or being disposed of within their borders can be challenged in the courts under the constitutional principles of intergovernmental immunity or federal preemption or under the commerce clause.[15] However, in the case of low-level wastes, recent legislation explicitly permits regional groupings of states to restrict as of January 1, 1986 the use of disposal facilities operated under a compact approved by the Congress to the disposal of waste generated within the region.

Among the most important nongovernmental actors in the nuclear waste policy arena are the numerous environmental protection and antinuclear special interest and lobbying groups. Many of these are well staffed with technically competent and legally and politically sophisticated activists who keep a watchful eye on federal activities, lobby the Congress and state governments, participate in NRC proceedings, and take full advantage of opportunities to challenge federal actions in the courts. For many of the active members of these organizations, the level of distrust and suspicion of government agencies and of most of their activities in the nuclear waste area would be difficult to overestimate. To many of these groups, of course, the nuclear waste issue and the evident lack of significant government progress toward opening a disposal facility for high-level waste, serve as a convenient and powerful argument and source of leverage in their efforts to prevent the expansion of commercial nuclear power or to shut down existing plants. They promote proposals to link nuclear power siting decisions to the nuclear waste issue in order to inhibit nuclear power growth, to provide incentives for stronger government action to deal with nuclear waste, or both. While all of the environmental groups would like to see nuclear waste stored and disposed of in a safe manner and believe their activities are assisting in achieving that goal, some of the antinuclear groups would prefer that the nuclear waste issue not be solved too quickly, lest this useful lever in their fight against nuclear power might disappear.

Industry groups, particularly the nuclear power plant suppliers and electric utility industry, also have a major interest in the nuclear waste problem. They correctly perceive that the absence of operating disposal facilities and the erosion of public confidence in the government's waste management and disposal programs have been detrimental to their efforts to retain public acceptance of nuclear power technology. They actively lobby the Congress to seek legislation that they believe would help solve their short-run spent-fuel storage problem and would improve the federal waste disposal program. They also work with DOE to help improve its programs and with regulatory agencies to seek protection of their interests. In these activities, the industry's approach used to be to urge programmatic haste, placing less emphasis on the need for thoroughness, programmatic redundancy, and due process than have environmentalists. Increasingly, however, some members of this interest group have come to believe that the government cannot make progress in the area of waste disposal without the support or acquiescence of the general public and those interest groups who are willing and able to erect debilitating political or legal barriers. They have begun to show as much concern for the technical quality, management efficiency, and institutional success of the DOE program as for its putative schedules.

The last nongovernmental group worth identifying as relevant to the nuclear waste issue is the scientific community. Many scientists, of course, support environmentalists or industry in pursuing their particular brand of advocacy. Others serve as consultants to or independent critics of DOE's, EPA's, or NRC's programmatic activities. Technical reports from agency contractors abound and are available for and sometimes purposely subjected to technical review by competent scientists, including those outside the community of government contracting personnel. The technical community is by no means united on most issues in the nuclear waste area. However, because of the technical complexities and uncertainties intrinsic to the subject, the pronouncements of notable scientific organizations or individuals in the nongovernment technical community do play a significant role in the molding of public perceptions of both the technical issues and the quality of government programs.

The National Academy of Sciences has traditionally had a special role through its standing Committee on Radioactive Waste Management and its ad hoc panels. Under contract, it supplies technical review and advice to DOE and NRC on a regular basis. However, the academy's record in this area is mixed. Some of its critics argue that in the past, academy reviews have done the program more harm than good.[16]

THE IRG PROCESS

Energy, nonproliferation, and environmental protection were all high-priority issue areas for the Carter administration, and nuclear power generally and nuclear waste in particular sat at their intersection. In the early policy reviews conducted between January and April 1977, the nuclear waste problem was recognized to be a serious threat to the continuing use of nuclear power, closely related to some aspects of nonproliferation policy, and a serious environmental problem requiring urgent attention. In the first National Energy Plan of April 1977, a task force to review the entire waste management program was called for under the direction of the assistant to the president for energy, James Schlesinger.

Despite some efforts at the staff level to launch this task force, it did not materialize immediately. Only after the Department of Energy came into being in October 1977 did an effort begin to reassess the existing policy and approach toward nuclear waste disposal. In December, DOE launched an internal study. Significantly, it was placed under the direction of John Deutch, director of the Office of Energy Research, not under the assistant secretary for energy technology who was responsible for the nuclear waste programs. The study's purpose was to help the new department develop its own view on how to deal with the nuclear waste problem and to prepare the way for a subsequent interagency effort. The report of the DOE study, the so-called Deutch report[17], served as a basis for the interagency review, and many of its conclusions and recommendations were adopted by that larger effort. The interagency study was the policy reevaluation, referred to earlier, that culminated with President Carter's message to the Congress of February 12, 1980 and tried to set a new course for the government's radioactive waste programs.

By the time DOE's internal study was launched, foreign policy considerations had already led to one major policy change. The evolution of the administration's nonproliferation policy led to a decision, announced on April 7, 1977, to defer commercial reprocessing indefinitely and to reorient high-level waste disposal programs to be capable of accepting either spent fuel or waste from reprocessing.[18] The administration recognized that this policy would exacerbate an already forseeable spent-fuel storage problem for some utilities that had counted on moving their spent fuel to reprocessing plants and were limited in the extent to which they could readily expand existing storage. The administration also wanted authority to provide certain countries with a near-term alternative to reprocessing as a way of dealing with spent-fuel accumulation. Initially, the administration intended to seek authority from the Congress solely for this latter purpose. However, the DOE decided it should help domestic utilities out of their spent-fuel storage problem and argued that Congress would not agree to acceptance of foreign spent fuel unless this was coupled with a government offer to take spent fuel from domestic utilities. The president was persuaded, and on October 18, 1977 DOE announced an intention to seek statutory authority from the Congress for the federal government to accept spent fuel from domestic utilities and limited amounts of foreign spent fuel in cases where doing so would serve the nonproliferation interests of the United States.[19] This policy was in place by the time the waste management review began, although implementing legislation was not actually sent to the Congress until late February 1979.[20]

During the initial DOE phase of the review, several of the strongest critics of the existing nuclear waste program from within and outside the government were able to communicate their concerns directly to the task force. Thus was established the important precedent of obtaining input from the widest possible range of views and perspectives. The product of the DOE study was a draft report[21] published in February 1978 that was given wide public distribution.

DOE's intention from the beginning of its in-house study was to move on to a second, administration-wide phase. This was in large measure the result of a perception that DOE's credibility—as the successor to the AEC and ERDA—was so low in this area and its ability to escape from past history was so constrained that it was simply not capable on its own of adequately reconstituting the nuclear waste management program. Therefore, shortly following the publication of the Deutch report, on the recommendation of the secretary of energy and with the concurrence of senior White House advisors, the president created an interagency task force, called the Interagency Review Group (IRG), chaired by the Department of Energy. Its purpose was "to formulate recommendations for establishment of an administration policy with respect to long-term management of nuclear waste, and supporting programs to implement this policy."[22] The IRG was comprised of policy-level officials of thirteen executive branch departments and agencies, including all those discussed in the previous section plus several—NASA, ACDA, and the Department of Commerce—whose interest in the subject was more specialized and parochial. The NRC, although not officially a member, was an active participant in much of the work. An IRG Executive Committee was constituted consisting of DOE, the Executive Office units, and later also DOI and the State Department.

The work of the IRG was carried out with great intensity. The DOE made clear that it believed nuclear waste to be one of its highest priority issues. It allocated extensive staff resources to the effort and, in addition, the chairman of the IRG, John Deutch, approached the subject with vigor and determination. He did not shirk extensive personal involvement or calling together the IRG or executive committee principals whenever matters arose that required their attention. At least until the final stages, he drove the process forward with a sense of urgency and commitment. The IRG's analytical tasks were assigned to six subgroups—high-level and transuranic waste, spent fuel, defense wastes, institutional issues, international issues, and transportation—each of which produced a report that was published. These fed into the work of the IRG Executive Committee which was responsible for writing the overall report of the IRG.

The work of the Subgroup on Alternative Technology Strategies for the Disposal of High Level and Transuranic Wastes is particularly noteworthy. As chairman of this group, OSTP set itself the task of doing a thorough technical analysis of the status of scientific and technical knowledge relevant to six different disposal approaches. To write the most important of these, that on mined repositories, a group of technical specialists was assembled from within and outside the government.[23] Numerous drafts were prepared and distributed for technical review, several quite widely, before official publication in draft form in October 1978. A special Technical Advisory Committee was constituted to review this technical paper and derivative policy analyses. It met some half-dozen times to review draft documents. Because of the diversity of the drafting team and the extent of this technical review, the final technical document was a unique contribution to the understanding of the status of knowledge relevant to geologic disposal at that time. It served as the basis for a paper produced by OSTP that identified and analyzed a series of policy options for the high-level and transuranic waste disposal programs.[24] The major thrust and content of this policy paper were subsequently adopted by the IRG in its own report.

The bureaucratic process by which the IRG report was written is of some interest in that it illustrates the proposition that progress can best be made on an issue as complex and emotionally charged as nuclear waste management through an institutional mechanism that permits active participation by diverse interests. Initially, DOE assumed responsibility as chairman for drafting the IRG's report. When its first product was ill-received by the Executive Committee, an interagency, staff-level drafting committee was created. Participating was a staff member from each of the Department of the Interior, the Office of Science and Technology Policy, the Council of Environmental Quality, and the Department of Energy. The latter acted as chairman. This committee, working under the direction of the policy-level members of the Executive Committee, wrote the IRG's draft report, reviewed the public comments, and produced the IRG responses to the public comment included in the final report. This procedure guaranteed that a range of bureaucratic, philosophic, and policy perspectives were represented and reflected in the documents. It also provided a mechanism for contentious issues to be dealt with and resolved or at least to be clarified before being confronted at the policy level.

There was considerable interest in the nuclear waste issue within the Congress during the period of the IRG, and the IRG tried to keep the Congress informed about its activities. Informal meetings were held with staff of relevant committees and with some of the most influential senators and representatives. Hearings were convened by several congressional committees focusing on technical, intergovernmental, and of course budgetary aspects of the government's nuclear waste program, and efforts were made by IRG members to be responsive, including by making available draft documents. However, the IRG did not try to bring the Congress or its members with most influence on the nuclear waste issue into the policy discussions. The Congress did not participate in the consensus-building process, and therefore neither was party to nor felt bound by the IRG's approach and recommendations. In retrospect, the absence of congressional power centers from the IRG process was one of its major failings.

However, involvement of the Congress would not have been easy. The constitutionally mandated separation of powers often makes members of Congress reluctant to become too closely associated with and therefore co-opted by executive branch proposals, particularly on controversial issues. They frequently prefer to respond to proposals only after they are formally presented. More important, because of the fragmentation of authority on this issue among the many congressional committees and the very significant differences in policy orientation among these committees, Congress could not have been adequately represented in the IRG process without expanding the number of participants beyond the point at which working groups would have remained manageable. Perhaps ways could have been found to make the Congress really participate in the policy-formulation process, but whether an internal congressional consensus and a congressional-executive consensus could thereby have been forged and at what procedural costs is difficult to guess.

The IRG recognized two reasons why obtaining input from interest groups and the public was critically important. First, it believed that its policy recommendations would not be useful or capable of being implemented unless they commanded broad support. Second, it believed that the legitimacy of the outcome and willingness of the public to accept it depended in large measure on the legitimacy of the process itself and on giving the public a chance to participate and be heard. In short, the IRG sought both to accommodate its policies to external reality and to draw relevant interest groups into the process in the hope that because of their involvement, they would be more likely to support the policy outcomes.

The result of the commitment to public participation was a series of public hearings in different cities across the country, meetings in Washington with interest groups, the availability of all documents for public review and solicitation of public comments on them, and a general openness and willingness to discuss issues and tentative policy positions with all parties interested in engaging in the dialogue. When the IRG draft report[25] was published in October 1978, 15,000 copies were distributed and comments were actively sought.

Some 3,300 written comments were received. These were screened and sorted by DOE personnel. All comments from government officials, offices or agencies, from private sector, environmental, or other special-interest organizations, from technical specialists, or from any unaffiliated individuals that discussed issues at length or raised new questions were passed on to the interagency drafting committee. These were perhaps a third to a half of all the comments. The drafting committee read all the comments it received and produced an integrated summary of their contents. The rest were brief statements, the substantive content of which was contained elsewhere, the repetitious results of letter-writing campaigns, or addressed to the issue of nuclear power rather than nuclear waste.

The review of the public comments, of course, led to the reopening of internal IRG discussions on many of the most difficult policy questions. To some extent, IRG members had been willing to let the draft report be published without it necessarily reflecting their own views precisely on every point. This was in the interest of moving the process forward and of maintaining the appearance of consensus on as many issues as possible and in the knowledge that issues could be reopened following (in fact would be forced open by) the public-comment period. Indeed, the draft report reflected disagreements among IRG members on only a few, albeit very important, points. The revised IRG report was published in March 1979.[26] It included the text of the original draft report, the drafting committee's summary of public comment on each section, and an IRG response to these comments. These reponses frequently involved extensions and revisions of previous findings and policy recommendations which resulted directly from the public-comment process. They also displayed more visibly the continuing disagreements among the various members of the administration through dissenting comments that appeared with the introductory formula, "Some agencies believe."

One important objective of the IRG, understood and shared by all members of the Executive Committee, was to move the waste management issue out of the narrow purview of DOE and into the broader context of the full government and the president. It was for this reason that DOE requested OSTP to take, and the latter accepted, the responsibility to lead the technical review and policy analysis effort on high-level and transuranic wastes. Such tasks would not normally be undertaken by OSTP or any Executive Office unit, but given the important of the issue and the absence of an acceptable alternative, OSTP reluctantly agreed.

In addition, the president's unique role as national leader was recognized to be an important asset to be employed in the effort to regain credibility for the government and to build public confidence in its ability to deal with the issue. The president, therefore, had to become personally involved. The IRG envisioned a public statement or message on nuclear waste that would put the president's own imprimatur on the IRG's recommendations and exploit the ability of his office to move the debate to a new, more constructive level. Of course, the policy issues that the IRG member agencies could not resolve among themselves also had to be presented to the president for arbitration. Surprisingly, there were only two: What to do about the Waste Isolation Pilot Plant (WIPP); and the extent of the redundancy that should be built into the repository site characterization program.

In the end, getting to the president turned out not to be easy. Drafting of an appropriate issues paper was extraordinarily difficult, even after publication of the final IRG report. Agreement could not be reached on what to say to the president or how to say it. Although only a few issues were left unresolved, they were very complex, important, and difficult to articulate clearly and succinctly. Agency positions were in some instances uncertain, as was the outcome of congressional action on the controversial WIPP project. Responsibility for drafting the presidential decision memorandum was moved from one place to another as first DOE and the Domestic Policy Staff, then the interagency drafting committee, now chaired by the OSTP representative, and then OMB failed to produce acceptable versions. The effort dragged on for months and was lengthened by the fact that all participants also had many other responsibilities (some related to nuclear power or energy generally, and some not) to discharge at the same time.

Not until early June 1979 was a paper that all agencies found acceptable finalized and ready to be sent forward to the president. Then other matters intervened: the Tokyo Summit of industrialized countries, the Vienna Summit to sign the SALT treaty, the Iran crisis, the recurrent efforts by the president to formulate a salable and aggressive energy policy, and finally the cabinet shuffle of July and vacations of August. The paper finally reached the president's desk in late August. Even then, progress was slow. First, he sent it back, asking for further analysis on the issue of program diversity. After that matter was resolved, the losers on the WIPP issue exercised their right to ask for reconsideration, resulting in further delay.

Drafting the presidential documents—the message to Congress, an accompanying fact sheet, an executive order, and instructions to the agencies—began before the last issue was settled. Again, leadership in drafting started with DOE and then shifted, this time to an interagency staff group comprised of DOE, OMB, OSTP, and CEQ representatives, and finally to OSTP. DOI also produced a version early on, some aspects of which were preserved. Substantial differences in view existed about the proper structure and emphasis of the documents. In many instances, individual sentences or phrases became subjects of heated and extended argument as draft after draft was circulated to the Executive Committee for comment. The presidential message did not finally emerge until February 12, 1980, more than two years after the internal DOE study began.

In retrospect, the long delays were perhaps inevitable because of the complexity of the issues and the continuing disagreements among the various members of the administration. Although largely masked in the draft IRG report, as the process came nearer the end and the president himself was to become involved in articulating a final policy, the disagreements came increasingly visible and made drafting very difficult. The role of OSTP became, as it had so often before, one of brokering, bargaining, seeking common ground, and ultimately arbitrating among the various parties.

Beyond this intrinsic difficulty, however, and the intervening of other demands on people's time, including the president's own time, there were other factors that contributed to delay. First, the president's political advisors in the White House were reluctant to have the president announce a nuclear waste policy soon after the Three Mile Island nuclear accident, which occurred just weeks after the IRG final report was published. Second, DOE with the concurrence of the Domestic Policy Staff and other White House political advisors wanted time to try to work out an arrangement with Congress that would permit the WIPP project to go forward in a way that the president would accept and that would avoid a confrontation with powerful forces in Congress. Even after the immediate impact of Three Mile Island had receded and the impossibility of getting Congress to agree to DOE's preferred disposition for WIPP became clear, a third reason seems to have become important: the president's political advisors were reluctant to announce a decision on the WIPP issue that was certain to displease Senator Frank Church of Idaho during a period when his help, as chairman of the Senate Foreign Relations Committee, was essential to achieve Senate approval of the SALT treaty.

RESULTS OF THE IRG

During the lengthy period before the presidential message was issued, the impact of the IRG was already being felt, particularly in DOE's program. Between the publication of the draft IRG report in October 1978 and President Carter's message in February 1980, two budgets were presented to Congress. These were reviewed by DOE and Executive Office personnel to ensure consistency with the IRG's recommendations. DOE's high-level waste program was broadened and reoriented, and many other actions recommended by the IRG were also implemented. However, far from everything could be done. The recommendations relating to institutional relationships were more difficult to implement without the president's explicit endorsement, and less progress was made in this area. One notable exception was DOE's relationships with states. These did evolve in directions endorsed by the IRG. Other agencies acted to a lesser extent than DOE to implement IRG recommendations. The EPA, for example, took virtually no action.

Even after President Carter's February 1980 message to Congress, implementation of his program was by no means complete. One problem was the continuing divisions on the nuclear waste issue in the Congress. Although the president's policy was strongly acclaimed in some quarters, it was severely criticized or even ignored in others. The leaders of the authorizing committees in both houses were among the strongest critics, and they managed through budget cuts and statutory language to inhibit or prevent implementation of several central components of the policy.

Equally serious was the failure of the White House itself to follow through. The president's message to Congress necessarily addressed only the most important policy issues. It settled the two questions on which the IRG had been split, and in some important areas explicitly diverged from the IRG recommendation. Of course, much of the details could not be included in the brief message, and these were either put in an accompanying fact sheet or left for the detailed implementing instructions intended to be sent by the president to the heads of executive branch agencies. However, although these instructions were mentioned in the fact sheet, and despite strong efforts from some quarters within the Executive Office, the Domestic Policy Staff could not be persuaded to have the president issue them. The result was ambiguity as to whether President Carter really accepted all the philosophy and subsidiary policy recommendations of the IRG. Even the status of issues addressed only in the fact sheet was questionable because it was not given very wide distribution, and subsequently the DOE preferred to ignore it. In addition, no procedure was put in place to monitor and enforce the scores of specific implementing actions that the operating agencies were supposed to and had agreed to take. Many of these were in fact never taken. The desirability of having an Executive Office coordinator in the months immediately following the issuance of the president's message was widely recognized, but there was no obvious candidate for the task and Executive Office units were reluctant to give up any of their authority. Therefore, a coordinator was never appointed.

The IRG devoted most of its attention to the disposal of high-level and transuranic wastes in the belief that these were the most significant categories of waste and involved the most difficult technical, political, and institutional problems. As already indicated, the subgroup led by OSTP did a thorough examination of the status of scientific and technical knowledge relevant to several approaches for disposal of such wastes and analyzed a set of alternative programmatic approaches. The judgment that disposal in mined geological repositories is the approach most likely to pay off in the near term was reaffirmed by the IRG and subsequently endorsed by the president. However, the IRG's statement about the current state of knowledge relevant to repositories and the work yet remaining to be done before one could be safely constructed was more cautious than previous government statements on these subjects. The need to examine the total repository system, including waste form, repository design features, local geology, regional hydrology, and regional tectonics in an integrated manner was emphasized. Different views still exist on how best to conduct a geologic screening program based on this so-called systems approach. For example, the Geological Survey would prefer to look for promising sites within regions with particular hydrologic characteristics. The DOE is proceeding with a program organized for the most part by rock type. However, the systems approach to characterizing particular potential sites has now been widely accepted and is a significant departure from the previous conceptual framework for dealing with geologic disposal which focused predominantly and relatively independently on waste form and disposal medium.

The IRG's discussions of risk assessment and of decision making in the face of continuing uncertainty are particularly noteworthy. Recognizing that no methodological alternatives to analytical modeling exist for assessing the risks of repository disposal, the IRG emphasized both the need to rely on such modeling and the intrinsic problems of doing so. Models are currently being developed by the NRC for use in licensing and by DOE for site characterization, and have been employed by EPA in formulating its high-level waste standard. An extensive examination of the current status of modeling methodology was undertaken and reported in the technical paper prepared by OSTP. The importance of sensitivity analysis and of considering the full repository system and the absolute requirement of conducting site–specific risk assessments at locations actually proposed for repositories were strongly emphasized. The fact that uncertainty is reduced at each successive step of site characterization and of the development and operation of repositories was also stressed and resulted in the proposal for stepwise, incremental decision making. That some residual risk would always remain was also explicitly recognized. The inevitable conclusion was to acknowledge that a social judgment about the acceptability of a particular level of risk would ultimately be required. Technical evaluation alone would be insufficient.

Consistent with these views about risk assessment, the philosophies of technical and programmatic conservatism and of proceeding toward a repository by means of a step-by-step approach, originally proposed by the DOE task force, were affirmed and strengthened. In a very significant departure from the dominant historical practice, the IRG proposed introducing greater diversity into the earth sciences laboratory R&D activities, the site characterization program, and the plans for in situ testing and data acquisition. As a result, DOE's waste form R&D program was expanded to include new technical approaches and its high-level waste disposal program was revised and expanded. A broader range of geologic environments is being examined and a more systematic site selection process has been adopted. In response to these changes in DOE's program, NRC's own preparations for reviewing a DOE license application have expanded and become focused on a broad set of geologic possibilities. The IRG did not explicitly examine the question of whether in situ examination should be conducted at sites before they are deemed ready for consideration in an NRC licensing proceeding. Rather, it simply accepted the precedent assumption that sinking a shaft would follow NRC construction authorization. Subsequently, however, the NRC reconsidered this question. Consistent with the IRG's philosophy of conservatism and its proposal of a step-by-step approach, and in part because of strong urging from the Geological Survey, NRC has now required that in situ testing at depth be part of site characterization for DOE's proposed repository site and alternatives and that it therefore be conducted prior to NRC reviewing a license application.[27]

The IRG also addressed the question of intergenerational trade-offs and concluded not only that "the responsibility for establishing a waste management program shall not be deferred to future generations," but also that the waste management system "should not depend on the long-term stability of or operation of social or governmental institutions for the security of waste isolation after disposal."[28] In other words, each generation should bear the total cost of the disposal of its own nuclear waste, and the risk to succeeding generations should be so low that they need be neither concerned with nor burdened by it. Because natural radioactive decay processes render nuclear waste gradually less hazardous over time and therefore also less expensive to dispose of, this philosophy of allocating disposal costs to the generation producing the waste can certainly be challenged. The disposal task could be accomplished more cheaply, more reliably, and with less occupational hazard in the future. The IRG's view, it is probably fair to say, derived as much from a desire to alter the public perception—fostered by lengthy delay and previous proposals for long-term retrievable surface storage—that the government was simply trying to avoid dealing with the difficult problem of disposal as from a thoughtful balancing of intergenerational costs, risks, and benefits.

Because of the importance to U.S. nonproliferation policy of the contention that reprocessing of commercial spent fuel is not essential to safe waste disposal and because earlier government statements to that effect had been made in the absence of complete technical foundation, the IRG, through the technical group at OSTP, addressed this question with some care.[29] Although no new technical analysis of this issue was conducted for the IRG, it tried to be accurate in characterizing the current state of knowledge and to place the question in its proper context. It also tried consciously to influence the discussions of this matter in other countries and internationally. The IRG stressed that when a repository is viewed as a total system, compatibility of the waste form to the host rock and environment becomes very important. Less was known about the chemical properties of spent fuel and its interaction with potential host rocks than was known for some waste forms resulting from reprocessing. In addition, some characteristics of spent fuel, such as the contained gases, were known to require special attention. The IRG acknowledged these things. However, it also stressed that compatibility can be achieved either by choosing a waste form to match a particular host rock and environment or by choosing a host rock and environment and designing engineered features of the repository that are suitable for a particular waste form. Combinations of these approaches are also possible. The IRG recognized that there may be somewhat greater difficulty in finding appropriate host rocks and environments for spent-fuel disposal than for disposal of reprocessing waste, which can be chemically and physically tailored. However, based on current knowledge, the IRG found no technical reason why appropriate sites should not be obtainable. The view that spent fuel could be an acceptable waste form for geologic disposal has now also been endorsed by the international community, as represented by the February 1980 final report of the International Nuclear Fuel Cycle Evaluation (INFCE).[30]

In recognition of the importance and complexity of the remaining outstanding technical questions, the IRG sought ways to improve the DOE's ability to obtain and utilize additional scientific and technical information and to employ non-DOE technical expertise. Greater use was proposed of the U.S. Geological Survey and the Bureau of Mines of the Department of the Interior. These organizations are the primary sources of earth sciences expertise in the government, including hydrogeology, geochemistry, and rock mechanics. In fact, interactions between DOE and the Geological Survey have increased, and DOE now transfers both funds and personnel positions to the survey. The two agencies produced an earth sciences technical plan[31] for the high-level waste disposal program in April 1980 that lays out an earth sciences program designed to address issues that remain unresolved. The survey's participation in the earth science component of DOE's repository program is extensive. However, the survey has been reluctant to mount a significant program of its own. At this writing, its long-planned initiative to identify candidate repository sites in the basin and range province of the western United States remains in abeyance. DOE never pursued a relationship with the Bureau of Mines, and the latter also let the prospect die. New advisory mechanisms were proposed to provide the advice of the nation's best earth scientists to the DOE's waste management programs. Scientific and technological advisory committees were assembled to advise. DOE's contractors, but none was created to assist the program's managers at DOE headquarters.

Prior to the IRG, DOE had expected to proceed to licensing of the first high-level waste repository after having found one qualified site. However, site selection is now expected to await the availability of several qualified sites in diverse geological environments and with a variety of potential host rocks. Such an approach was acknowledged to cost more and to require more time, at least as calculated on program planning schedules. However, the IRG believed that a comparison among sites, each of which was believed to be suitable, would provide added assurance of safety and was quite possibly essential to satisfy the legal requirement of NEPA to examine reasonable alternatives. Redundancy would also result in an overall program with more robustness and resiliency in the face of continuing technical challenge and political crosscurrents. The IRG further believed that states would be more likely to acquiesce in the early stages of site investigation if they knew that sites in their state would later be compared with other sites elsewhere before choice. Finally, the IRG argued that despite apparent delays on planning charts prepared under the optimistic assumption that programs are permitted continually to move forward, the actual time required for success might well be shorter with redundancy and site comparison. This followed from the expectation that steady progress was more likely through a diversified approach than by trying to focus early on the first available site.

The extent of redundancy—as measured by the number of sites that should be available before the first site is selected by DOE and a construction authorization request docketed with NRC—was a matter of disagreement among IRG members, and as previously indicated, was one of two issues settled by the president. The DOE, supported by the State Department and ACDA, urged a comparison among only two to three suitable sites prior to choice. All other IRG members expressing a preference except DOI, and including all the president's relevant White House advisors, argued for a broader intercomparison of four to five sites. DOI preferred not to identify a number, but to let the degree of comparison emerge out of subsequent program planning.

When this question was presented to the president in primarily numerical terms, he asked, not surprisingly, why three to four sites was not a sensible compromise. This forced a clarification which focused more directly on the real issue at stake. These included judgments about how much programmatic redundancy and deliberateness were needed to regain public confidence and whether the existing DOE program was likely to produce an adequate diversity of qualified sites or whether selection should be delayed until the expanded program had time to yield results. In addition, there existed differing views about whether a procedure that the planning schedules said would take longer might not actually turn out to be faster because it would be more acceptable. The DOE argued that the greater redundancy would not be helpful, that delay was unnecessary, and that the appearance of delay would be harmful. Others believed that the program as then organized was not likely to yield adequate sites and that DOE and OMB needed a strong incentive to effect programmatic expansion. They also argued that reducing the apparent time pressures in the short run would provide more opportunity to improve the institutional aspects of the waste disposal problem before siting choices must be faced, and therefore that the probability of ultimate success would be greater. In the end, the president agreed with the majority view despite an evident reluctance to accept a programmatic approach that could be characterized by critics as simply further delay.

Relying on this revised program philosophy, the DOE estimated the earliest possible date for site selection at 1987.[32] Initial repository operations were projected for 1997. These dates were based on the assumption that, as previously indicated, NRC licensing would precede any *in situ* activity at a proposed repository site. Now that the NRC has indicated that *in situ* testing at depth is an essential component of characterization for both the proposed site of the repository and alternatives, these dates should be slipped by three to four years.

The NRC's final rule[33] for the procedural aspects of licensing a high-level waste repository also requires that DOE complete characterization, including *in situ* testing, at a minimum of three sites in two geologic media, at least one of which is not salt, prior to submitting a license application. Although less conservative and involving less redundancy than President Carter's policy, this NRC requirement incorporates its philosophy and general approach.

In situ experiments with heaters, simulated waste, and perhaps live waste can provide information about chemical interactions between the waste or packaging materials and potential host rocks and about heat transfer and conductivity. According to NRC's requirements, such experiments would be initiated during the characterization of a particular potential repository site and then continued or even expanded during the early years of repository operation. Data of a generic nature can be obtained from separate R&D facilities at locations not intended to evolve into repositories. If all the waste emplaced in such facilities were removed, they would not require licensing. DOE has three such projects underway for *in situ* testing with nonradioactive materials. One of these, a facility in basalt on the Hanford reservation, is scheduled to receive live waste in 1982. A granite test facility on the Nevada test site already contains radioactive materials. The third, in a Louisiana salt dome, does not involve actual waste. DOE is also participating in similar work conducted by other countries, notably the examination of crystalline rock at depth near Stripa, Sweden.[34] Because of the high cost of separate R&D facilities and the requirement for *in situ* characterization at actual proposed repository sites, the cost-effectiveness of such facilities has been challenged. For example, DOE's request for approval of another R&D facility to allow emplacement of radioactive material in salt was turned down several times by OMB during the 1970s.

The IRG believed that much could be learned both institutionally and technically by building a larger but still modest-sized licensed disposal facility ahead of the first full-scale repository. The term "demonstration" has frequently been used to characterize such a facility. However, because the adequacy of a repository is very site specific and can only be verified after the passage of many thousands of years, the technology cannot be demonstrated in the usual sense. The IRG therefore believed this term was misleading. It invented the name "intermediate scale facility" (ISF) instead. ISFs as conceived by the IRG would be licensed, highly instrumented, and designed to permit retrievability of waste if necessary, and would contain no more than several hundred to a thousand waste canisters or spent-fuel assemblies.

The IRG agreed that although an ISF is not an essential component of the waste disposal program (because the same learning could take place in the early phases of the first full-scale repository), if an appropriate opportunity to build an ISF were to arise on a schedule significantly prior to the opening of the first full-scale, high-level waste repository, it should be taken. To the DOE, the Los Medanos site near Carlsbad, New Mexico, where characterization was well-advanced and a repository for defense transuranic waste—the WIPP project—had been planned, represented an appropriate opportunity. Other IRG members who expressed a view on this matter, again with the exception of the State Department and ACDA who shared DOE's desire to make early visible progress, did not believe that an intermediate scale facility should be sited at the first available location. They preferred to carry over the philosophy of site comparison and wait until a choice could be made from among two or three suitable sites. This issue was addressed by the president as part of the general question of how to treat the proposed WIPP facility.

The more important issue with respect to WIPP was whether to proceed or cancel the project. Congress had already authorized it and funds had been appropriated. This project originated from a promise made by the AEC to Idaho's governor and senators that transuranic waste produced within the defense programs and stored at the Idaho Nuclear Engineering Laboratory would be removed for disposal at an early date. No one associated with the IRG challenged the analysis indicating that transuranic waste could be disposed of most efficiently within a high-level waste repository and that no urgency exists from the perspective of health and safety to remove this waste from Idaho before the first such repository was available. Therefore, the project's real purposes were to demonstrate the ability of the government to make progress on waste disposal and to live up, albeit belatedly, to this earlier political promise.

The facts that transuranic wastes are not as heat generating as high-level waste and that under existing law a facility intended for disposal of only transuranic wastes from the defense program would not be licensed by NRC meant that the project as initially conceived would actually contribute little to the main high-level waste program. Recognizing this, DOE had proposed, in the Deutch report, adding an intermediate scale facility (then still called a demonstration facility), which necessarily would be licensed, to the WIPP project. DOE also proposed a change in the law to require a license for the disposal of all transuranic waste, whether deriving from civil or defense activities, just as the disposal of all high-level waste independent of origin is now required to be licensed. Following publication of the Deutch report, DOE tried hard to have these proposals accepted by the Congress. So determined was DOE to go forward with the revised project, that the department, in its role as chairman of the IRG, at first tried to rule WIPP out of bounds for IRG consideration. The Executive Office units, however, refused to accept this constraint, believing that what happened to the WIPP project was an absolutely fundamental issue with respect to the president's approach to nuclear waste management.

The governor of the state of New Mexico, while not endorsing the project, was not opposed to DOE's continued planning for the facility, including the associated intermediate-scale facility, so long as it would undergo independent licensing review and the state could exercise a concurrence right as previously promised by the secretary of energy. It was also reluctant to oppose a project judged to be important by the armed services committees that had authorizing jurisdiction. These committees also had authority over other federal activities in the state that are critical to its economy. The mayor of Carlsbad and the elected state officials from the region supported the project because they foresaw employment and other benefits for the area. The local state representative and senator chaired the relevant committees in the state legislature and were able to obtain the legislature's support, or at least acquiescence.

In Congress, the armed services committees of both houses wanted to proceed with the defense facility and opposed any complications that would delay progress or that would bring additional defense waste activities under NRC licensing authority. They opposed the colocation of an intermediate scale facility with its associated disposal of commercial spent fuel and requirement for licensing. These views were shared by the leadership and majorities of the Senate Energy and Natural Resources Committee and both appropriation committees, and they were supported by full congressional action.

All IRG members, including DOE, recognized that the project might turn out not to be feasible because of questions concerning the suitability of the proposed site, or because of potential opposition that might develop from the state or Congress. The presence of known hydrocarbon and potash reserves at the site led some on the IRG and many nongovernment observers to be skeptical about the site's suitability. However, based on information then available, the DOE believed the site to be suitable and issued both a site characterization report and a draft environmental impact statement on the project while the IRG's work was in progress. Subsequently, DOE acknowledged that more information was needed on the regional hydrology, and appropriate work was begun.

DOE, the State Department, and ACDA felt that the project (initially defined in IRG discussions as a licensed transuranic repository with colocated intermediate scale facility) should be permitted to go forward through the NEPA, congressional, and regulatory review process and not be stopped preemptively by the president. It was, they pointed out, the only opportunity for significant near-term progress in the repository program, and they believed such progress to be essential to the program's overall credibility. Others preferred to cancel the project, protect the site at which tens of millions of dollars had been invested, and delay the first disposal facility for transuranic waste and an ISF until more sites were available from which to choose. They believed that the likelihood that the proposed project at the specified site would win acceptance was very small. They argued that if WIPP went forward now, skeptics would claim that, depsite the IRG, previous policies still predominated. This, they feared, could undermine the credibility of the entire IRG process. Even worse, they feared that unless the administration actually cancelled the project, it would end up being unlicensed and therefore subject to political decision made by Congress and the DOE, not necessarily based on objective technical evaluation.

By the time the WIPP issue went to the president for formal decision, the attempt to build support in the Congress for DOE's preferred position was recognized to have failed. The armed services committees were unwilling to let the project disappear from the defense waste program or to have it licensed, and none of the civilian authorizing committees would sponsor the project as a licensed transuranic waste repository with an intermediate scale facility. Reportedly, the president would have preferred the DOE approach if it had been feasible, but under the circumstances, he decided to cancel the project, protect the prospective site, and delay siting a transuranic waste facility and intermediate scale facility until a comparison of sites could be made. The DOE, however, was not willing to accept this result. Claiming that Secretary Duncan, who had just recently replaced Secretary Schlesinger, had not had a chance to review the issue before it went to the president, DOE took the issue back a second time. This time DOE argued for acceptance of the project as authorized; that is, as an unlicensed repository for defense transuranic waste without an intermediate scale facility. Again, the president opted for cancellation.

According to the president's decision, the WIPP project was to be cancelled and the Los Medanos site to be treated simply as another candidate location for a future high-level waste repository. The strategy for implementation was to seek a recision of appropriated FY 1980 funds and zero the WIPP line item in the FY 1981 defense waste budget. Necessary funds for continued site work would be included in the commercial budget that is not reviewed by the armed services committees. This strategy failed, however. Congress denied the recision request and wrote into the FY 1981 budget authorization for defense waste activities a requirement for DOE to proceed with the project as an unlicensed R&D facility for defense high level waste including the disposal of defense transuranic waste.

The Los Medanos site is federal land under the jurisdiction of the Bureau of Land Management of the Department of the Interior. Therefore, prior to sinking a shaft and conducting *in situ* work, DOI must agree to access to the land by DOE under the provisions of the Federal Land Policy and Management Act (FLPMA). A fallback strategy for implementing President Carter's decision on WIPP, therefore, was for DOI to refuse to take the necessary action or simply to delay. Under FLPMA, an environmental impact statement is necessary prior to DOI's formal consideration of a land withdrawal. The DOI under the Carter administration considered DOE's WIPP environmental impact statement inadequate for purposes of FLPMA and requested a new document that included a more specific discussion of the proposed uses of the land. This view was expressed formally in a letter dated January 19, 1981 from the DOI solicitor to DOE's Assistant Secretary for Environment.

In the meantime, however, DOE pressed ahead with preparations for the WIPP project as authorized by Congress. In October 1980 a final environmental impact statement was issued and on January 22, 1981 it issued a record of decision to proceed with the project. The impediments at DOI were quickly removed by the Reagan administration. The letter of the previous solicitor was withdrawn and on April 2, 1981, a cooperative agreement between the two departments, was signed authorizing DOE to proceed.

President Carter's message significantly stepped back from the IRG's earlier support for going ahead with a licensed ISF at an appropriate time in the future, by not even mentioning the concept. The accompanying White House fact sheet's statement that an ISF would have substantial benefits reflected a view held within the government at that time only by the fact sheet's primary drafter, OSTP. DOE did not believe the licensing exercise was worth the cost and effort, and prefered to rely on testing during the early years of operation of a facility intended to be a full-scale repository to obtain the technical benefits. Nonetheless, the concept is not dead. Responding to Congressional pressure and its own strong desire to make visible progress as soon as possible, DOE has now included in its program plans a test and evaluation facility to be constructed at one of the sites where *in situ* characterization is expected to be completed by 1985. Unlike intermediate-scale facilities, as conceived by the IRG, this test and evaluation facility would not be licensed by NRC. This is an unfortunate feature of a concept which, while it will be controversial, appears to this observer to be fundamentally sound. The negative aspects of the absence of licensing will be ameliorated to some degree if DOE builds this facility at a site it would like to consider for a future full scale repository. This is the current plan and will require that all planning documents and technical reports be given to NRC for their review and comment. NRC might well set specific requirements that DOE must meet.

For reasons suggested earlier, institutional aspects of waste management were recognized by the IRG to be at least as difficult and important as the technical aspects, and important changes were proposed for the program's institutional context. The most significant involved trying to design new relationships between the federal government and state and local governments in the siting of high-level waste repositories. As already mentioned, these relationships had seriously deteriorated in many states; and federal activities, including the very early stages of site evaluation, were being impeded or even prohibited. The IRG sought mechanisms to institutionalize cooperation between states and the federal government and to encourage states to be active and—most important—willing participants so long as federal programs protected the health and safety of their citizens. One approach that the president endorsed was to provide federal funds and technical assistance to states and other jurisdictions to facilitate their working with DOE and ultimately to permit their informed participation in review and licensing proceedings. DOE has already begun working in this way with several states and the results so far seem beneficial to all parties.

The most important and difficult aspect of the federal/state relationship is the allocation of decision-making authority. As previously mentioned, states do not now have authority in the high-level waste area, although, as a practical matter, they have strongly influenced the DOE activities within their borders, including in some instances causing all program activities to cease. ERDA and

subsequently DOE had decided, as a matter of policy, that the federal government should not site a repository in a state without its agreement.[35] The IRG
believed not only that the practical political power of states could not be
ignored, but also that states had a legitimate right to be a major participant in
decision making with possible effects on the health and safety of their residents. Therefore, it believed that procedures should be designed to accommodate the states' interests. The two situations the IRG most sought to avoid were
a state's preventing any site investigation work from going forward or its
acquiescing in expensive and time-consuming site-specific activities only to veto
a proposed project late in its planning.

The relationship proposed by the IRG to address the states' interests was
named consultation and concurrence. This implied the establishment of dialogue and "the development of cooperative relationship between states and
relevant federal agencies during program planning and the site identification
and characterization programs, . . . through the identification of specific sites,
the joint decision on a facility, any subsequent licensing process and through
the entire period of operation and decommissioning." The state would have the
opportunity "to participate in all activities at all points throughout the course
of the activity and, if it deems appropriate, to prevent the continuance of
federal activities.[36] The IRG hoped that if the states were made equal partners
in DOE activities and as a matter of policy given the right to suspend federal
activities, they would be more willing both to let site characterization work go
forward and to agree to construction of a facility if a site were found suitable.

The IRG's concept of consultation and concurrence turned out, however, to
be both controversial and confusing. The confusion came from the fact that
although the IRG explicitly disavowed the term "veto" to describe its preference for a state's role, it nonetheless proposed that states have the authority
that the word "veto" implies, namely, the right to stop a federal activity or
project if, after extensive consultation and attempts to reach agreement, it
chose to do so. The IRG preferred to emphasize the process of cooperation,
bargaining, and accommodation between DOE and states, which it called
consultation, and to deemphasize the possibility of nonconcurrence. The IRG
hoped that so long as consultation worked well, the likelihood of nonconcurrence would be small. Moreover, the IRG believed that focusing attention and
discussion on the possibility of nonconcurrence was unlikely to foster an
atmosphere conducive to establishing constructive consultation relationships
and getting on with the needed work. However, the manner by which the IRG
chose to explain its intention in its draft report did lead to rather considerable
confusion. Despite clarification in the final report, the confusion in the minds
of many interested parties was not alleviated.

The controversy derived from the existence of rather different views within
the Congress and among students of both the waste management issue and the
American political process about how much authority the states really have

now and how much they should be given. On the latter question, for example, differing levels of authority would be granted to states by the various draft bills submitted to the ninety-sixth Congress prior to the February 1980 message, but the most widely supported ones did not go as far as the IRG had proposed. A number of conferences on consultation and concurrence were held following the release of the IRG final report, sponsored by DOE's primary contractor and by others. At none of these, however, did consensus emerge on the question of what would be the best role of the states in the siting of high-level waste repositories.[37]

In light of this confusion and controversy, the president in his February 12 message made only a hedged endorsement of the consultation and concurrence concept. He used the term, but defined it only as giving to host states "a continuing role in federal decision making on the siting, design and construction of a high-level waste repository." He did not specify what the nature of or authority implied by that role should be. He also stressed the need for states to act as partners and accept their share of the national responsibility.Clearly, and probably wisely, he decided to reserve judgment on the precise nature of the optimum federal state relationship until further discussions had taken place within the national community interested in nuclear waste management.

In order to serve as a forum for discussion and further development of the consultation and concurrence concept and for other purposes, President Carter created—by executive order on the day his message to Congress was released—a new organization called the State Planning Council on Radioactive Waste Management. This council consisted of elected state, local, and tribal representatives. It was intended to ensure that state, local and tribal perspectives were provided to federal agencies and reflected in their program. It was originally suggested as a council of governors by the National Governors Association, but was broadened by the IRG with NGA approval. The executive order gave the council an eighteen-month life, and the Carter administration unsuccessfully sought legislation that would make it permanent. The council was advisory to the president and the secretary of energy. It was directed to address questions with respect to siting criteria, transportation, interim storage of high-level waste, the adequacy of the low-level waste disposal system, the adequacy of DOE's R&D and site characterization programs, and procedures for implementing consultation and concurrence.

The IRG anticipated that the council would help DOE to understand and take into account state perspectives. Because it might be less parochial than any individual state, it was expected also to share DOE's view that steady progress must be made. Its recommendations would be easier for states to adopt than federal policy, however, because the council would itself be a representative of states' interests. The council was not intended to share authority with the federal government, to substitute for authority being given to individual states, or to interfere with the consultation and concurrence process between the federal government and particular states.

The IRG also sought ways to diffuse and distribute the political opposition anticipated to be directed at particular governors and state legislatures when their states become candidates for siting of a high-level waste repository. It therefore adopted a concept of regional siting of repositories that was proposed by the DOE task force and whose origins can be traced back to ERDA. This concept was strongly endorsed by state and local officials in order to avoid the reality or appearance of any particular state becoming the single nuclear garbage dump for the nation. The IRG and subsequently President Carter adopted the concept, and it has been incorporated into DOE planning. The idea did generate opposition from those who feared that regionality might result in geographic or political boundaries taking precedence over geologic suitability in siting decisions. This was not at all the intention, and the IRG and the president carefully qualified their endorsement of the concept by specifying that regionality should be adopted "insofar as technical considerations related to public health and safety permit."[38]

The IRG and President Carter recognized that serious problems existed in the area of low-level waste disposal and that urgent action was necessary. Of the six original commercial low-level burial grounds, three are closed. The remaining three, in Barnwell, South Carolina, Beatty, Nevada, and Richland, Washington, are not optimally distributed geographically and are physically incapable of taking all the nation's low-level waste for many more years. Moreover, the state governments and citizens of the three states have also become concerned about the adequacy of both commercial practices and regulatory oversight of packaging and shipping. The western sites were temporarily closed in the fall of 1979 in order to force a reduction in the number of violations of shipping and packaging rules. South Carolina announced that the volume of waste it would accept—in recent years, as high as 80 percent of the country's total low-level waste—would be cut in half. In addition to these immediate problems, the funds available to provide adequate perpetual care at some existing and inoperative commercial low-level burial sites are inadequate.

The IRG proposed that the federal government should assume responsibility for developing and implementing a national strategy for low-level waste disposal. In addition, the IRG recommended that states be given the option to retain management control of and responsibility for existing commercial low-level waste sites or to transfer such control and responsibility to the federal government. The states, however, preferred that they be in charge of the low-level waste problem and expressed this view forcefully through the National Governors Association and the National Conference of State Legislators. President Carter, therefore, did not endorse the IRG's earlier recommendation for possible federal takeover, but rather supported the actions already taken by governors, particularly the three with operating sites in their states, to promote solution of this problem at the state level.

This approach was embodied in law in December, 1980 with the passage of the Low-level Radioactive Waste Policy Act. This act makes individual states responsible for disposal of their low-level waste and urges cooperation on a regional basis through the mechanism of interstate compacts. Discussions are underway in most regions of the country on the formation of such compacts and in some cases on locating new disposal sites. Although achieving regional agreements and especially the siting of new facilities will be difficult and probably take several years to complete, a process is now underway that is likely to achieve those ends. Still remaining is the issue of providing adequate care of existing inoperative facilities. A federal role is probably essential here, and legislation will be needed to authorize it.

The IRG also proposed an increase in the level of R&D focused on new and improved technologies for disposal of low-level waste. This has occurred. DOE's appropriations for low-level waste technology were $12.8 million in FY 1981, all of it within the defense waste component of the budget, and the Geological Survey's own program in this area had $2.3 million. Volume-reduction techniques and alternatives to shallow land burial are being pursued, and work is underway relevant to site selection and operational criteria. Little thought has been given, however, to how to transfer new ideas and technology to the states and private sector entities likely to be responsible for siting and operating new facilities. The general assumption seems to be that new facilities will be shallow land burials just like the old ones, but this may not be the best approach. Alternatives include using deep mines, an approach already employed in some other countries.

The IRG gave considerable attention to the problem of decontamination and decommissioning (D&D) of the many existing surplus government facilities. As a general rule, unrestricted land use was recommended as the ultimate objective of D&D. Institutional controls should be in place for only a limited period and should not be relied upon to provide long-term protection of people and the environment. However, for some existing sites or facilities where total decontamination is not possible at a reasonable cost or perhaps at any cost, long-term institutional control may be required. DOE's strategy for proceeding with a long-term D&D program incorporates this philosophy. Legislation will be necessary to provide DOE with authority to do the required work at some sites. The IRG also proposed that the D&D expenses should be included in the estimated cost of new government nuclear projects at the time of authorization, but no action has been taken to implement this approach.

In the area of uranium mill tailings, the IRG broke no new ground because the necessary programs were already in place and regulations in preparation. In 1973 work began to clean up structures in Grant Junction, Colorado, where mill tailings had been utilized as building materials. This program is 75 percent funded by DOE and 25 percent by the state. As of April 1980 about one-third of the structures were completed. The 1980 schedule called for program completion in 1987. Under the Uranium Mill Tailings Radiation Control Act of 1978, DOE also is responsible for cleaning up some twenty-five inactive mill tailing sites at which uranium for national defense programs was processed before they were abandoned. Work is scheduled to begin in 1983 pending completion of EPA cleanup and disposal standards with which DOE must comply. Interim standards for cleanup of open lands and buildings were published in April 1980,[39] and proposed disposal standards were published in January 1981.[40]

For active uranium mills, NRC or agreement states have regulatory oversight. NRC's final rule for Uranium Mill Licensing Requirements was published in October 1980.[41] Agreement states that exercise regulatory authority over uranium mills must employ standards at least as stringent as those beginning in November 1981. However, at this writing several agreement states are doubtful they can meet that deadline and the entire regulation has been put in question by a suit brought by the American Mining Congress. EPA standards, with which NRC's must ultimately conform, do not yet exist, but, pending policy level review within the agency, a draft is expected to be published in late 1981.

For reasons given earlier, the IRG believed that the absence of standards and regulatory procedures in many areas of nuclear waste management, the slow schedules on which EPA and NRC were working to produce these, and the order in which they seemed likely to emerge constituted a serious problem. Although it proposed a number of actions to deal with the problem, most of them were not implemented. The IRG proposed and President Carter directed in his message that the regulatory schedules be accelerated. However, the regulatory agencies took little action in response. NRC did accelerate its schedule for producing a low-level waste standard in the absence of which badly needed, new low-level waste disposal facilities cannot be opened, and draft regulations did appear in 1981. The IRG recommended that the agencies be provided with resources to permit acceleration. As the numbers provided earlier indicate, this happened for NRC but not for EPA. The IRG also obtained EPA's agreement to produce a paper that would give advance guidance to NRC and DOE concerning the possible form of its standards and other matters that these agencies would like to have at an early date. Such a paper was not produced and no plans to produce it now exist. In addition, the IRG proposed that EPA and NRC produce a memorandum of understanding that would address schedules, methodologies, and the exercise of overlapping jurisdictions. No action has been taken. Although some draft and final standards have been issued by both agencies in the last two years, the prolonged schedule for the rest still remains a serious problem both for ongoing program activities and for public perceptions of the government's ability and willingness to deal with radioactive waste.

The IRG considered the question of whether NRC licensing authority should be extended to waste activities now exempt. Consistent with the proposal that the federal government offer to assume the responsibility for low-level waste disposals, the IRG proposed that any new government facilities for disposal of commercial, low-level waste should be licensed. However, subsequent action giving states the responsibility for low-level waste disposal made this proposal moot. The issue of defense waste activities was much more complicated. The principle of equal protection and equal regulatory oversight for equivalent waste, independent of its source, was very attractive at least for new sites that could be built to be licensable and had been adopted by Congress in the limited area of disposal of defense high-level waste. Some existing storage and disposal sites would require large expenditures before the NRC would approve them. Others probably could not ever be made licensable because the knowledge of what had been buried in the past is so incomplete. However, powerful forces in the Congress—particularly the armed services committees that have jurisdiction over the defense waste programs—were known to oppose any further NRC licensing of defense waste activities. Some IRG members felt these forces might oppose other reforms the IRG was recommending if the president sought an expanded NRC role. The resolution of this dilemma selected by the IRG and the president was a limited proposal to extend NRC licensing to disposal of defense transuranic waste. This was consistent with the proposal to cancel WIPP and dispose of the defense transuranic (TRU) waste in the first facility for high-level waste disposal.

Before the president's message, the NRC published a staff report on licensing requested by Congress.[42] It recommended extension of NRC licensing authority to cover all new DOE facilities for disposal of TRU waste and non-defense low-level waste, thereby endorsing the IRG proposal. It also recommended testing the feasibility of establishing a consultative relationship between NRC and DOE with respect to the latter's waste-management activities not covered by NRC's licensing authority or to be covered by the recommended extension. Although internally the IRG had agreed to reexamine the licensing issue when this NRC report was published, it never did so.

As things turned out, forces in the Congress opposed to expansion of licensing took the lead in this area. As already indicated, the armed services committees succeeded in writing statutory language that prohibits the licensing of defense transuranic waste disposed in WIPP. Thus, even the IRG's limited proposal will not be adopted unless this congressional attitude is reversed. That seems unlikely. Indeed, strong pressure has been exerted by some members of the armed services committees to amend current law to remove from NRC oversight the disposal of defense high-level waste.

The IRG and the president recognized the importance of public and interest group participation in all aspects of waste management program planning. The IRG believed that because of the degree of public participation in its own activities, a greater willingness to cooperate had been generated within relevant government agencies and nongovernment groups than had been the case for many years. Mechanisms were proposed to expand the opportunities for continuing public participation, including the provision by the government of technical and financial assistance as needed to permit informed public input to programs and decisions. President Carter explicitly addressed this subject in his message, but again in the absence of implementing instructions, the agencies received no directive to act and no procedure was put in place to ensure that they did so. Little change in fact has occurred, except in DOE.

During the IRG process, the suggestions were made repeatedly by many nongovernment observers that responsibility for nuclear waste management should be taken from DOE and put in a new independent organization and that overall coordination and oversight should be vested in either a continuation of the IRG or within the Executive Office of the President. However, the IRG believed that another reorganization to create a new agency would be very debilitating to the program, and preferred to seek ways for DOE to regain public confidence and do a better job. The IRG also rejected the notion that either an interagency committee or an EOP agency could really be effective as a coordinator or overseer body on a continuing basis. Rather, it believed, DOE must have sufficient authority to run the program and the many existing EOP oversight mechanisms should be strengthened. In fact, however, EOP oversight was not strengthened within the Carter administration. No White House coordinator was named and most of the EOP staff involved in the IRG were eager after issuance of the president's message to turn to other issues.

The IRG did propose, and the president directed, that a systematic nuclear waste planning process be instituted through the vehicle of a biennial comprehensive National Plan for Nuclear Waste Management. This plan was to be submitted for public review in draft form and then revised, based on public comments. The plan was to address NEPA implementation, low-level waste, uranium mill tailings, decontamination and decommissioning of government facilities, and earth sciences and waste form R&D and site qualification for the high-level waste program. It was also intended to contain updated summaries of the status of scientific and technical knowledge relevant to disposal of each type of waste. A draft of the first version of the plan was released for early review by state and tribal officials and the Congress in January 1981.[43] DOE originally intended that it would subsequently be revised and reissued for public review, but the new administration has not announced its intentions.

The draft plan did not include summaries of the status of scientific and technical knowledge. However, a separate document was published by DOE in October 1980 intended to fill that role for high-level waste.[44] This document is not the product of an extended drafting process relying heavily on non-DOE personnel, as was its predecessor prepared by the IRG. Rather, it is based largely on that predecessor evaluation, with additions borrowed heavily from DOE's statement of position submitted to the NRC Confidence Rulemaking.[45]

Prior to publication, the document was reviewed by the U.S. Geological Survey, DOE and DOE contractor personnel, and by a handful of nongovernment specialists.

The final noteworthy proposal by the IRG and adopted by the president in principle was a new attitude toward program schedules and target dates. In the past, a date associated with the expected opening of a repository has taken on the character of political commitment and a criterion by which progress was expected reasonably to be judged. The IRG recognized the importance of planning schedules and target dates for imposing discipline and providing program incentives. However, it also recognized that the time required to accomplish the many tasks contributing to opening a repository is very uncertain and depends in large part on events and circumstances beyond the government's control. The IRG therefore rejected the notion that planning schedules should be considered political commitments or that slippage should be interpreted as programmatic failure.

Implementation of this view, of course, is very difficult. Summary statements, charts, and tables which become widely used to characterize a program can rarely include appropriate caveats and hedges. For example, even the president's message referred—albeit with some hedges—to specific dates by which site selection and initial repository operations were expected to occur. In addition, politicians and program managers alike are driven to make firm promises in order to demonstrate that they are in control and are competent managers. The pressure on DOE from the NRC confidence proceeding is particularly important in this respect. Perhaps inevitably, therefore, DOE has tended to slip back into previous practice. Although the repository schedule it uses is called a reference schedule, many of DOE's publications lack much recognition of the uncertainties. They adopt, instead, a highly optimistic tone. The quality of some recent DOE documents also suggests they were released more to meet particular schedules than because they were actually ready.

In the area of spent-fuel management prior to the opening of a repository or a decision to reprocess, the IRG only endorsed policies already in place that had emerged out of nonproliferation considerations. Parallel to the IRG's activities, DOE endeavored without success to achieve congressional support for these policies though passage of the administration's spent-fuel legislation. DOE's judgment was probably correct that the Congress would not grant authority for the government to accept small amounts of foreign spent fuel in support of nonproliferation objectives without also putting DOE into the business of storing domestic spent fuel. Indeed, there was very little enthusiasm in Congress for the foreign component of the policy even with the linkage. To DOE, the linkage was also a strategy for seeking acquiescence for a government role in domestic spent-fuel storage from those inside and outside the government who would support the foreign spent-fuel offer. This did not work either. The environmental interest groups generally supported the foreign spent-fuel offer derived from nonproliferation policy so long as it remained limited, but opposed the government's becoming involved in domestic spent-fuel storage.

The government involvement turned out to be the most significant objection to the policy. DOE argued that even if utilities expanded existing at-reactor storage capacity to the maximum extent possible and transferred spent fuel from full pools to others within the utility's system, full core reserve would be lost for a few operating reactors by 1983. DOE further argued that the private sector could not provide the needed away-from-reactor capacity in time to prevent loss of full core reserve. It proposed that the government take over and expand one or more of the existing storage pools at nonoperating reprocessing plants to provide initial capacity and later, if needed, to build a new facility. The case was not sufficiently persuasive to the Congress. The estimates of need were very uncertain, and the argument that the utility industry was unable to deal with the problem on its own in time was rejected by many.[46] In addition, many wanted to avoid having to move the same spent fuel twice, first to an away-from-reactor storage facility and later to a reprocessing plant or repository. President Carter renewed his call for passage of the spent-fuel legislation in his nuclear waste message; and the legislation, absent the foreign spent-fuel component, did have strong support from powerful congressional figures, particularly in the Senate Energy and Natural Resources Committee. Nonetheless, despite strong efforts in the waning days of the ninety-sixth Congress, no legislation on this subject was forthcoming. Indeed, congressional supporters and industry agreed to delay consideration of the issue in order to increase the chances of getting a bill passed dealing with high-level waste disposal.

The IRG articulated a principle of financing waste disposal by which the cost would be paid by the generator of the waste and borne by the beneficiary of the activity in which it was generated. This approach is already applied to the commercial burial of low-level waste, which proceeds on a fee-for-services basis, and to the care and disposal of defense wastes, which are financed from the federal budget. It has yet to be institutionalized for commercial high-level waste. The mechanism the IRG proposed to do so was strongly influenced by the administration's spent fuel policy. The one-time fee that utilities would pay under this policy occurring at the time the government took title to spent fuel, would cover the cost of: all government R&D and site characterization programs, past and future; repository construction, operation, and decomissioning; and the cost of any government strorage involved. Collecting this fee early would both provide funds not tied to annual budget appropriations for the conduct of DOE's repository program and allow utilities to include the cost of waste disposal in their rate computations. Because the policy presumed that spent fuel would remain at reactors to the maximum extent possible, utilities were originally going to be permitted to exercise the option of giving up title to fuel stored at reactors. In that case, they would also pay a fixed fee that would not include the cost of interim storage and would therefore be lower. This option was later dropped, with the result that the one-time fee proposal—even if put into practice—would have applied to only a small part of the total spent fuel. The establishment of a special trust fund to hold the money was proposed in the administration's spent-fuel bill. However, the attempt to give the secretary of energy disbursement authority independent of congressional oversight was controversial. In the absence of congressional action, of course, implementation has not occurred.

THE FUTURE

The Reagan administration and the ninety-seventh Congress were left a legacy in the radioactive waste area that, in most respects, was much better than the one their predecessors inherited. Yet, this legacy was still deficient in many ways and had a particular orientation deriving from the work of the IRG. As the new administration assesses its own approach to this issue, it must face some major philosophic, programmatic, and budgetary issues. This final section will review the major areas requiring attention and will explore the fundamental issues needing resolution in each case.

High-level waste disposal remains the most difficult and politically charged area. The most critical programmatic issue is the extent to which the philosophy of site comparison prior to choice and the associated diversified geologic investigation program will be maintained. As indicated earlier, the NRC has required that DOE include in the environmental report supporting a license application site characterization data, including results of *in situ* testing from at least three sites in at least two media, one of which must not be salt. The commission also indicated that "in light of the significance of the decision selecting a site for a repository, the Commission fully expects the DOE to submit a wider range of alternatives than the minimum required here.[47] Unless Congress acts to countermand this rule or a subsequent commission amends it, the philosophy of comparison now has the force of law, albeit in a somewhat weaker form than President Carter proposed. The questions for DOE and the administration, therefore, are whether they will comply only with the minimum requirement of the NRC rule, or go further as NRC "fully expects," and whether, whatever the strategy, sufficient funds will be provided to the program to let it go forward at a reasonable pace.

Given President Reagan's efforts to reduce federal spending, budgetary pressure might force minimal compliance. Budgetary savings might be made by reducing the number of sites examined simultaneously. In the next few years, savings of some tens of millions of dollars, out of a total commercial high-level waste management program of about $200 million per year, might be achieved in this way. However, the savings in later years could be greater because characterization of several sites is scheduled to reach the stage where shafts must be sunk and *in situ* testing performed simultaneously. In fact, however, this does not appear to be happening. The Reagan budget gave the high-level waste repository program an increase in fiscal year 1982 over fiscal year 1981, and DOE is anticipating an accelerated site characterization program including the sinking of exploratory shafts at three sites in fiscal year 1983. This might then be followed by a test and evaluation facility at one of these sites. In its program description accompanying the revised fiscal year 1982 budget, DOE did not say whether it would adopt a minimum compliance approach and plan to apply for a license after three sites had been characterized. Instead, it continued to hold open that question for later decision. Consistent with this approach, DOE's record of decision on the generic environmental impact statement on the management of commercial generated high level waste, addressed only the limited issue of selecting mined repositories as the disposal option choice. No statement was made about the notice or diversity of the program directed at siting and developing such repositories.[48]

The extent of necessary interactions with state officials and dealings with internal state politics could also be reduced by narrowing the scope of the program. In the limit, geological investigations could be restricted to DOE reservations. But serious questions still exist with respect to the adequacy of sites at both Hanford and the Nevada Test Site, and even the NRC's minimum requirement can probably not be met if such a strategy were adopted. Moreover, narrowing the program might conceivably have the opposite effect by making the smaller number of targeted states less cooperative.

According to the present reference schedule, two or three years could be saved by proceeding after a comparison is made between only three sites instead of five, but this estimate is very uncertain. The actual savings could be more if work goes quickly at some sites and slowly at others, or it could be less or even negative if the institutional problems of working with states are exacerbated by narrowing the program. The administration and Congress will have to decide whether the savings in expenditures of dollar, effort and time as measured on reference schedules appear significant enough to reduce the degree of programmatic insurance and risk a loss of state and government and public confidence. To this observer the trade-off appears to favor maintenance at least for planning purposes of a diversified program based on broad comparison prior to site selection.

Another important issue for the high-level waste disposal program is the extent to which states and Indian tribes that are potential hosts for a repository, or that are adjacent and potentially affected, will be granted rights and authority. There was considerable, although not universal, support in the last Congress for giving potential host states a right to object formally to the Congress to a DOE repository siting proposal, although not a right to veto such a proposal. The dominant view seemed to be that one house of Congress must agree with the objecting state in order to stop the project from going forward. Legislation dealing with this issue was not enacted, however.

One area of disagreement was when in the process a state's opportunity for objection should come. State officials would prefer that it precede NRC review of a DOE application for construction authorization, arguing that Congress would be very unlikely to sustain a state's objection once NRC had determined the proposal to be consistent with public health and safety. Others argue that Congress should consider primarily nontechnical matters and should have before it, at the time it reviews a state objection, the NRC's technical judgment about the adequacy of a site from the perspective of public health and safety.

As directed in the executive order creating it, the State Planning Council considered in great detail the question of what authority states and tribes should have in repository siting and how it should be exercised. The council agreed that states should not have veto authority. This is perhaps surprising for a group that represents state interests. It derived from a view that high-level waste disposal is a national problem and that states acting individually should not be able to prevent its solution. It also derived from a recognition that most state elected officials would find acquiescence in a proposal to site a repository in their state very difficult politically and, therefore, that they would rather be able to pass the ultimate responsibility and the blame on to others.

In its eighteen month existence, the council put together a package of proposals on how the federal/state/tribal interaction should work in the site characterization program and the site selection process.[49] This package is more detailed and complete than anything yet written into proposed legislation. It would require that states and tribes cooperate with DOE in the early stages of site characterization and urges written agreements between responsible federal agencies (certainly DOE, but perhaps also DOI and DOT) and states and tribes. It recognizes the existence of informal as well as formal procedures by which states can object to DOE actions. It would allow the state to object formally to Congress prior to NRC review of DOE's proposal and would require positive action by both houses of Congress to overrule a state or tribal objection based on certain specified criteria for objection. It would also give affected states and tribes equal standing with host states and tribes on issues of direct concern to them and an independent right to object. This completed package is now available to the Congress or DOE for incorporation in whole or in part in legislation or policy if they wish to do so.

The views of powerful interest groups on the issues of states rights and the extent of diversity and comparison in the repository program are worth noting. The environmental community strongly supports the approach of a highly diversified site investigation program and comparison among four to five sites in different geologic environments prior to selecting the first site for which to seek a license. It also favors the strongest possible role for state and tribal governments. Environmentalists believe such approaches would maximize the likelihood of protecting the public health and safety and provide adequate time and opportunity for the institutional problems among federal agencies and between federal agencies and states to be resolved. Increasing state, local, and tribal authority over repository siting decisions would also give the environmental groups greater opportunity to influence the decision-making process.

Most state and local officials also share the environmentalists' views on the preferred degree of diversity for the repository program. They believe that a broad approach would most completely protect their interests and partially relieve the inevitable political pressure they will be under to oppose DOE site-characterization activities within their borders.[50] There is, however, disagreement within this community whether the extent of state and tribal authority over siting decisions should include veto power. As indicated, the State Planning Council did not favor state or tribal veto, but this is not a universally shared view.

The nuclear industry is somewhat divided. Some of those in or associated with industry who follow nuclear waste issues believe that the waste problem can be readily solved if only the government would just get on with the job in an expeditious way. They oppose a requirement for comparison among sites prior to choosing the first repository site and oppose giving states or tribes any formal authority or rights. They support proposals such as that by Representative McCormack in the last Congress that would implement this approach, mandate near-term milestones for the high-level waste disposal program, and circumvent the usual, but lengthy processes and procedures.

Some observers in the industrial community have the view that a slower, more deliberate repository development program that pays more attention to institutional problems and builds a firmer scientific and technological base is the best way to proceed. Holders of this view believe that attempting to move more rapidly and without active participation in decision making by states and tribes is likely to result, not in progress, but in stalemate. They also see their long-term interest in doing what is necessary to build public confidence in the government's efforts. They would like to be able to argue that, as a direct result of government programs in place, the public can be confident that disposal facilities will in due time become available, and therefore that nuclear power should be allowed to proceed.

Several years ago, the first of these views strongly predominated in industry circles. However, a shift has taken place in the direction of the second view. This shift in industry opinion produced an interesting and powerful coalition of industry and environmental groups during the post-election session of the ninety-sixth Congress in late 1980. This coalition worked hard to achieve enactment of legislation that gave states limited rights and included a requirement for a diversified site characterization program, but with firm milestones specified and no requirement for site comparison prior to choice. As mentioned above, industry even agreed to put aside temporarily the away-from-reactor storage provisions it seeks in order to forge a coalition for this legislation addressing high-level waste disposal. The effort failed, but not over an issue about which these interest groups disagreed. Senator Jackson sought to exempt disposal of defense wastes from the bill's provisions giving states limited rights to object to federal repository siting decisions. The House conferees did not agree, and neither side would yield.

Whether support for this bill from industry and environmentalists represented tactical compromises adopted in order to encourage enactment of a bill at that time, or whether it reflected an actual change of view, remains to be tested during the ninety-seventh Congress. Industry might have been willing to accept limited state authority over siting decisions in order to get the procedures written into law before an administration feared to be more states-rights-oriented took office. Environmentalists might have been willing to give up on the requirement for site comparison in order to have legislation enacted before an administration anticipated to be unsympathetic to their perspective took office. They might also have hoped that the NRC's procedural rule or eventual court challenge under NEPA would resolve the issue of comparison in their favor. With a more conservative and pronuclear Congress and administration, and now that NRC's procedural rule is final, industry may seek to reinstitute a fast-track approach by legislation. The environmental community and their still-powerful supporters in the House are likely to resist such efforts vigorously and may themselves strive for a stronger state role.

One very possible result of the continuing disagreements over these fundamental issues is that legislation that lays out a strategy for high-level waste disposal and a philosophy for federal/state/tribal interaction may not be enacted, or it may be delayed in the present Congress. Because of this possibility, it is important to point out that generic legislation is actually not essential to continued progress. In its absence, the geologic investigation program would be guided by the requirements of the NRC's procedural rule, with the pace of activities determined by the annual budget process. The Department of Energy could, as a matter of policy, specify how it will interact with states and tribes and what role they will have in decision making about repository siting. DOE cannot grant states legal authority over siting decisions, but it can, as a matter of policy, decide to act only in accordance with a set of procedures that de facto give the states and tribes certain rights. Indeed, this is precisely the policy DOE adopted during the Carter administration. Such a DOE policy could of course be rather readily altered, and the only way that states and tribes could be provided certainty about the rules would be through legislation. But such certainty may not be essential for progress on federal/state/tribal relations and for program activity to go forward.

Whether, on balance, legislation is considered desirable will be judged by each interest group according to whether it expects the outcome to be better or worse from its perspective. Although environmentalists and state interests

might gain through legislation setting the rules for the exercise of state authority, they might lose if the same legislation overrules the NRC requirement for
site comparison through the NEPA process. Industry might achieve the mandated schedules it has sought and perhaps even countermand NRC's rule by
obtaining a NEPA exemption, but might end up with states having specific and
extensive authority to block progress.

How these issues will be resolved by the ninety-seventh Congress—or
whether they will be—depends importantly on the approach taken by the
Reagan administration, how hard the industry and environmentalist interest
groups decide to push their preferred outcomes, and what their relative power
in the new Congress turns out to be. In addition, of course, what Senator
Jackson decides to do about seeking special treatment for defense wastes and
whether industry and its Senate supporters will seek to recombine the repository program with spent-fuel storage in a legislative package will have important implications for the outcome and its timing.

One unfortunate aspect of most bills considered by Congress has been their
mandated schedules, a feature strongly supported by the industry lobbyists
and some environmentalists. Not only were the schedules contained in the
various bills almost certainly unattainable, but the pressure that such legislated
deadlines would put on the program is likely to be counterproductive. Clearly,
the Congress wants to and should encourage DOE to make progress, and the
department should be held accountable. But, as experience in the area of
standard setting by federal regulatory agencies has clearly shown, unreasonable mandated deadlines are not a constructive approach to achieving such
objectives. Moreover, by trying to mandate a schedule by legislation, industry
leaves itself open to those who will try to make successful performance
measured against that schedule a prerequisite to reactor licensing.

By not acting before August 1981 to extend its life, President Reagan let the
State Planning Council go out of existence. The council's record as an advisory
group representing the views of states, tribes, and localities to federal agencies
had been quite respectable. It managed to attract good staff and to create
useful networks of informed and interested experts from industry, academia,
environmental groups, and elsewhere to help with each of the issues it addressed. The council was a significant factor in achieving consensus in the low-
level waste area, provided useful input to NRC and DOT in their rule makings
relevant to high-level waste disposal and transportation, provided constructive
comments to DOE on drafts of the first National Plan, and has prepared a
useful approach to federal/state/tribal relations in the high-level waste disposal program. The council also acted occasionally as a helpful mediator
between particular states and DOE when the former have been unhappy about
actions taken by the latter. Whether, in its absence, the National Governors'
Association or the National Conference of State Legislatures will try to perform some of its functions or whether DOE will be open to and encourage such

action remains to be seen. DOE might feel it can come out better in its dealings with states if it faces them individually rather than collectively.

Whether or not there is legislation that addresses federal/state/tribal relations in the high-level waste program or another group fills the State Planning Council's place, the primary relationship will be that of day-to-day interactions between DOE or DOE contractor personnel and state officials, citizen groups, and the press. In recent years, these relationships have in general improved. The governor of Michigan, for example, agreed to discuss how DOE might recommence site investigation activity in that state. Agreements have been signed between DOE and Texas and Louisiana. But there is still much room for improvement in the ability of DOE and DOE contractor staff to work with states and tribes with understanding and sensitivity. The experience in New Mexico is particularly noteworthy because it will be scrutinized closely by other states seeking clues about how they can expect to be treated by DOE. After lengthy negotiations on a consultation and cooperation agreement mandated by Congress, DOE refused to sign the document if it was made legally enforceable and unless the state gave up its right to judicial review. DOE then proceeded to issue a record of decision on its environmental impact statement which said, "Construction of permanent surface and underground facilities will proceed on a phased basis."[51] The state believed that both good practice and NEPA required a more limited decision to proceed only with the conduct of the *in situ* exploratory program called Site and Preliminary Design Validation (SPDV), and a further review of the additional data before commiting to full-scale construction. On these and other issues New Mexico sued DOE and sought a preliminary injunction against drilling then underway at the site. A hearing was avoided in July only by DOE agreeing to sign a binding, enforceable agreement as the state had wanted, to pause following SPDV for a considered look at the new data and decision whether to construct the full facility, to resolve open land access issues, to help address anticipated socioeconomic impacts on the state and to seek funding to conduct additional tests requested by the state. It is unfortunate that legal action had to be threatened before DOE agreed to meet the state's requests. At least on paper, however, the stipulated agreement of July 1 seems to set a reasonable basis for federal-state relations and could well serve as a useful precedent for other cases.

Most states need to put in place organizations to interact with and monitor federal agencies and procedures to ensure internal communication and coordination within the state government. Many will also have to augment their own state personnel with academic or other experts to enable them to make independent technical evaluations and judgments about DOE's conclusions and recommendations.

Another area of federal/state/tribal relations that needs more attention is impact assistance. In many instances when the federal government opens a large facility, such as a military base, it provides the state and local governments with funds to compensate for the additional demand for local services. Sometimes, needed construction is financed in advance from the federal treasury. This approach seems quite attractive for repository siting and has been proposed by the State Planning Council. Indeed, the council has suggested that the extent of socioeconomic impacts and the amount of compensation be determined early in the process as a means of muting possible state and local opposition. This issue must be studied further and could usefully be included in any legislation on federal/state/tribal relations enacted by the ninety-seventh Congress.

DOE's latest estimates, indicate that no shortage of spent fuel storage capacity at reactors will occur until 1986 at the earliest,[52] and the Reagan Administration does not support federal government involvement in interim commercial spent fuel storage except for the funding of R&D. If the utility industry and its supporters in Congress accept that change in policy, the diversionary and ill-fated effort to bring the federal government into this business will end and industry can set about ensuring that adequate storage capacity will exist until a repository is available or reprocessing begins and catches up with the backlog. If industry and Congressional backers continue the fight for a federal away-from-reactor storage facility, the issue will remain open until Congress votes decisively. There can be little doubt at this point that the private sector can deal with the problem technically and, given clear signals from the government, will figure out how to do so financially. All the incentives push in that direction, and sufficient time is available.

Some segments in the industry hope for a federal takeover of the uncompleted reprocessing plant at Barnwell, South Carolina. If such a takeover occurred, the government could then complete that plant through an investment that no private company is likely to find attractive and could begin reprocessing operations. In the meantime, the pool could be used for away-from-reactor storage. Indeed, if the whole operation could be carried out under the guise of R&D or brought within the legal umbrella of the adjacent DOE Savannah River facility, it might even escape licensing. The current secretary of energy, who is also a former governor of South Carolina, favored this approach and sought funds in the revised FY 1982 budget to implement it. The president, however, disallowed the request and has now stated as his policy that reprocessing must be a private sector activity.[53] This policy seems likely to be sustained by the Congress if challenged.

In recent years, industry has argued that a government spent-fuel storage could be shepherded past the many political and regulatory obstacles more easily than each utility on its own could obtain approval to build additional capacity or permission for intra- or inter-utility spent-fuel transfers. There is also uncertainty whether any firm would want to or could raise sufficient capital to build a new private facility. Neither argument is totally convincing.

The demonstrable need for additional storage capacity, independent of uncertainties in future government policy about reprocessing and the repository program, and the high incentives for utilities to have an off-site solution to their spent-fuel problems ought to be translatable into a highly profitable private sector venture. Moreover, the intrinsic uncertainties and political nature of everything the government tries to do in the nuclear area, combined with the certainty that any efforts to site and license a large spent-fuel storage facility will attract the concerted opposition of every antinuclear group in the country, argue for a private sector and, to the extent possible, decentralized solution.

Persisting in pursuit of a federal solution to the spent-fuel storage problem is a risky strategy from the industry's perspective and may play directly into the hands of the antinuclear critics. If no action is taken until it is too late to avoid reactor shutdowns, the spent-fuel issue will indeed have turned out to have been an Achilles heel of the industry, as many critics now hope and believe.

The administration and the Congress also have an opportunity to implement fully for commercial nuclear power the widely supported philosophy that the generator of nuclear waste should pay the cost of disposal. Some sort of federal tax on nuclear-generated electricity that was passed through by utilities to their customers would be one of several possible approaches. It would have the advantage of immediately transferring the cost of waste disposal to consumers, and would therefore probably win support of environmental groups. It would also provide the government with a cash flow to help finance R&D, site characterization, and construction related to the repository program. This would be most useful in a period when strong downward pressure on the federal budget might result in funding cuts and substantial program delays. The disadvantage would be the political problems of imposing a new tax in an era when political trends are going mostly in the opposite direction.

In the area of interim care of defense wastes, some expensive actions, including completion of the transfer to new tanks of liquid wastes at Hanford and Savannah River, really must be completed in order to ensure continuing safe operations. Such work is underway and is likely to continue to command high priority. Other things, such as decontamination and decommissioning of surplus facilities and initiating construction of the planned Defense Waste Processing Facility at Savannah River, have been postponed for years for budgetary reasons. DOE has repeatedly sought and failed to receive administration support to seek construction funds for the latter facility. Given the exigencies of the budget, such deferrals are likely to continue. The argument can be made that no purpose is served by proceeding with waste processing facilities needed to prepare liquid wastes for movement and disposal on a schedule significantly faster than a repository will be available. Indeed, with adequate care and attention, public health and safety can be ensured for continued interim storage of defense high-level wastes for a long time. In addition, selection of the ultimate waste form for disposal probably should await knowledge of the geologic medium into which the waste will be placed. But the budgetary demands of the overall waste management program will only rise, especially when and if construction of the WIPP facility begins, and when shafts begin to be sunk at potential repository sites. There will never be a good time to make the large investments required to decommission facilities and to prepare the defense high-level waste program for disposal. If the country truly feels an obligation not to force future generations to deal with our wastes, a start must be made.

Budgetary constraints are likely to impact significantly, not only on the timing of actions to prepare defense high-level wastes for disposal, but also on how disposal should be done. Preparing Savannah River wastes for disposal off site in a repository will be very expensive, and preparing the Hanford wastes for geologic disposal will be extremely expensive. The question must be faced, therefore, of how to balance costs and benefits. The major benefit sought is public health and safety. There are financial costs and the costs of radiological impacts on workers. The current view is that for Savannah River and Idaho Falls, the balancing comes out in favor of off-site geological disposal. For Hanford, the case for geologic disposal is less certain. Permanent entombment in place might be better.

For the wastes at the nonoperating West Valley reprocessing plant, these questions of cost-benefit trade-off among options and timing of activities have been settled by Congress. The West Valley Demonstration Project Act calls for total cleanup of the site by DOE and a demonstration of solidification techniques that can be applied to other high-level liquid wastes, including those at defense sites. If budgetary pressures delay action at Savannah River, then West Valley, at which the political pressure for action is much stronger, may be the first waste processing facility built.

Budgetary pressures might also influence decisions about the WIPP project. All other considerations aside, WIPP will be a very expensive facility for disposal of a relatively benign form of nuclear waste which can be safely left where it is until disposal at low marginal cost in the first high-level waste repository. Recently, a different approach to transuranic waste disposal has been proposed that merits further investigation.[54] It would employ a deep surface depression made by an underground nuclear test at the Nevada Test Site—and, if feasible, would be much less expensive than WIPP. This location would not be appropriate for high-level waste disposal, but, if built without NRC licensing, the WIPP project also will contribute virtually nothing to the mainstream high-level waste program. For now the Reagan Administration appears determined to proceed with the project and the FY 1982 budget revisions added substantially to the request for it. Whether budgetary or policy considerations will lead to an altered view over time remains to be seen.

Despite passage of the Low-level Radioactive Waste Policy Act at the end of

1980, much remains to be done to implement its intentions. States, both individually and by regional groupings, must assess the quantities of low-level waste that must be dealt with in coming years and move toward creating compacts and siting facilities. This will not happen easily or rapidly, because opposition can be expected to almost any siting proposal. In the northwest and southeast, where facilities already exist, progress toward a regional compact has been made. But in other parts of the country, there will be greater difficulties. In the meantime, contingency plans must be formulated on a regional, statewide, or generator-specific basis for interim storage of low-level wastes. The act permits compacts to exclude waste from outside their member states beginning in January 1, 1986, setting this as a deadline by which new sites in other regions must be available or interim storage must be initiated.

There is clearly a role for federal technical and perhaps financial assistance as the states move in these directions and the DOE is preparing to provide it. Some state officials, including the State Planning Council, would like to see federal funds available to provide impact assistance to host communities and states. In addition, there is broad agreement that states should be given the option to transfer a decommissioned site and perhaps the existing closed sites prior to decommissioning to the federal government for perpetual care. Legislation would be required to authorize such transfer and impact assistance.

In several respects, the federal government must still get its own house in order. The absence of final NRC and EPA regulations in several areas, alluded to previously, bears reemphasizing. In the area of high-level waste repositories, NRC's procedural rule is final and the technical component is now out in draft form for public comment.[55] The opening of new low-level waste facilities requires completion of at least NRC's rulemaking. And, of course, EPA actions on which everything else depends are likely to be the last to arrive. Already NRC has been sued once, on its mill tailings regulations, on the grounds that its actions preceded those of EPA on which NRC must depend. The Reagan administration has made clear its belief that the federal government issues too many regulations, and might act to slow down the issuance of new ones. If, in doing so, the administration were to distinguish between regulation of private sector activity (which is presumably its major concern) and regulations needed for the conduct of the federal and state government's own programs, little or no effect might result on the nuclear waste program. At this writing, however, the administration's intentions and ability to make such distinctions remain unclear.

Interagency relationships in several areas where the IRG tried to make improvements still require attention. The relationship between DOE and the Bureau of Land Management on questions related to the use of public land are still not resolved. Nor has the role of the U.S. Geological Survey in DOE's program or as a possible consultant to NRC been fully clarified. A memorandum of understanding between DOE and DOI would still be useful. DOE and

DOT must still work out relationships over transportation issues that will be relevant in the siting of a high-level waste repository.

The internal structure of the high-level waste disposal component of the nuclear waste program also remains a problem. The shortage of personnel, especially with technical competence, the indirect reporting of contractors through DOE field offices, and the absence of centralized direction over the site investigation program, all dilute DOE's ability to manage, coordinate, and plan the program. The decision in late 1980 to have the primary contractor, Battelle Memorial Institute in Columbus, Ohio, report directly to headquarters is a step in the right direction, but will not solve the whole problem. DOE recognizes the need for more people and is trying to increase its staff in the commercial waste program. The Carter budget sought an increase of fourteen positions, mostly technical people to be resident in Columbus, near the lead contractor.

The extent of the problem at headquarters is illustrated by how drafting of the National Plan was handled. The fact that a contractor was hired for the purpose is perhaps surprising in itself, given the document's importance for the program's public image. Nonetheless, such an approach can work if the contractor is given sufficient guidance and help by program managers. But such was not the case in the area of interagency coordination where a contractor cannot be effective. The various drafts, including the one distributed in January 1981, suffer from this neglect.

The Congress, again led by the armed services committees, has a different concern about DOE's organization. By statute, Congress has directed that the defense waste program be separated from the civilian one and be placed under the programs of the assistant secretary for defense. This action, like the attempt to restrict states' roles in site selection for repositories to dispose of defense high-level waste and the insistence that the WIPP project proceed as an unlicensed defense facility, is another attempt to prevent the problems of commercial nuclear power and the uncertainties of expanded decision-making procedures from inhibiting the defense nuclear programs. The return to separation between defense and commercial programs, as in the days of the AEC, would have serious consequences, however. It would be quite disruptive in the short run and perhaps wasteful of resources because the two programs as now designed should interact very substantially. Perhaps more important, the public credibility of the defense waste program will likely plummet if the separation is made. Indeed, the very reason for making it—to proceed with greater independence and fewer delays—will be widely interpreted as equivalent to taking less care to protect public health and safety. This has been the result in the case of the WIPP project.

The location of the waste management program in DOE still remains controversial in several respects. As a result of a recommendation in the Deutch report, the nuclear waste program was moved from its former position subordinate to the nuclear energy development program into a position at the deputy-assistant-secretary level parallel to nuclear development under the assistant secretary for energy technology. When that assistant secretary position was abolished and an assistant secretary for nuclear energy was created, the waste program became part of his portfolio. Many observers of DOE have argued that this was an unfortunate location for the waste program. First, it makes the waste effort subordinate to the objective of promoting nuclear power—presumably the primary objective of the assistant secretary for nuclear energy—and risked the kinds of neglect or misdirection that resulted when this relationship prevailed in the AEC and ERDA. Second, and probably more significant under that arrangement, the head of the nuclear waste program was not a presidential appointee and was not at the policy level within the agency. Given the political sensitivity of the nuclear waste issue, the requirement that the program director interact with elected state officials, and the need for him to exert leadership among the various federal agencies, the director should have been a policy-level official who could speak with authority for the DOE. Both of these concerns, but especially the second, argued persuasively for elevating the director of the nuclear waste management office to the level of assistant secretary or its equivalent.

Of course, President Reagan's decision to fulfill his campaign commitment to abolish the DOE raises new questions. Many of the functions of the department must continue to be performed and, therefore, must be transferred elsewhere. Perhaps the nuclear waste program would be set up as a separate operating agency as has been proposed in the past.[56] Except for the short-term effects of still another major reshuffling, that would probably be a beneficial development. Whether the defense programs would be included remains to be seen. An alternative would be to move the defense waste and the nuclear weapons program together, perhaps to the Department of Defense.

The new administration will also have to decide how much effort to exert in continuing to rebuild public confidence in the waste disposal program and in obtaining public participation. Many people argue that the Carter administration placed too much emphasis in this area. They rightly point out that neither total consensus nor total societal acceptance of federal proposals in the area of nuclear waste management will ever be possible. As usual, when costs and risks are distributed within society, those who perceive themselves likely to receive adverse impacts must be expected to appeal to technical arguments, the political process, and legal means to prevent such an outcome. In addition, some social activists without direct interests at stake can be expected to intervene, arguing that proposed actions involve risks to the public health that are too high to be acceptable. Finally, some antinuclear groups will be loathe to see progress made toward nuclear waste disposal for fear of losing an effective weapon against nuclear power. For these reasons, the skeptics claim, efforts to build consensus or gain wide public confidence will not avoid controversy, but only cause delay. Better, they say or imply, to move more quickly, precipitate the inevitable confrontation, and seek to overwhelm the opponents through the exercise of political power. Parenthetically, but not surprisingly, many holders of this view also believe that their interests will predominate in any such political contest.

This view can be challenged on several fronts, although its analysis of the inability of public confidence-building and public participation efforts to avoid confrontation seems correct. The first challenge is an efficiency argument. The history of nuclear waste management—and many other government efforts directed at mitigating public health hazards—amply demonstrate that the government acting on its own or even in concert with a single set of private sector interests is by no means assured of acting in a way that will maximize the general public interest. A broader representation of perspectives and interests, although it will frequently complicate the decision process and delay programs, has a better chance of maximizing the general interest or, put differently, minimizing total costs and risks and maximizing benefits to society as a whole.

The second challenge is distributional. In the absence of a process that involved consensus building and compromise, the politically stronger side is more likely to prevail in decision making. Without a voice and procedural safeguards, the weaker party is without protection and is likely to suffer disproportionately by the proposed distribution of costs, risks, and benefits.

Third, the legitimacy of an outcome to a political or administrative process depends in large part on the perception of legitimacy of the process leading to that outcome. Participants or those who feel they have been adequately represented are likely to be co-opted by the very fact of participation and therefore to be more willing to accept an outcome even if they believe that they have lost. Without a willingness of losers to accept the legitimacy of a policy outcome, the many avenues of appeal and delay will be exercised, and progress becomes extremely difficult.

Fourth, diehard opponents, such as antinuclear groups who wish to exploit the nuclear waste issue as leverage against nuclear power, are unlikely to have decisive strength on their own. They can succeed only when they can attract sufficient allies who, although not sharing their extremist views, nonetheless agree with many of the criticisms they raise. Public participation and improvement in the public image of the government's nuclear waste management program, while unlikely to convert the extremists, can undercut their political power base and thereby undermine and greatly reduce their effectiveness.

For all of these reasons, the Carter administration's commitment to public participation and confidence building would appear to have been expedient and wise. The new administration would be ill-advised to reverse that commitment. It is to be hoped, therefore, that the public review, revision, and republication of the National Plan will be vigorously pursued and repeated in two years time. It is to be hoped that DOE will continue to build relationships with state and tribal officials and to become more open to criticism and open for public access to information. It is to be hoped that the technical advisory committees now helping DOE's contractors will continue and that a broader-based public advisory committee will still be created to advise policy-level, responsible officials. It is to be hoped that the administration will resist congressional attempts to separate the defense waste program from the commercial waste program, which, from the perspective of public relations, cuts exactly in the wrong direction.

Despite these actions that the administration can take, the most important determinant of public confidence in DOE's program in the near term is likely to be the outcome of the NRC confidence proceedings. Should the commission decide that it does not have a high degree of assurance that radioactive waste from nuclear power plants will be safely disposed of or that such disposal is likely to be available on a schedule much slower than DOE's reference schedule, the DOE program will be seriously undercut. In addition, the implications for further licensing of nuclear power plants could be profound. It is the latter factor that has generated intense interest in the proceeding and which transforms it from a purely technical to an intrinsically political one. Alternatively, if the commission decides that it does have a high degree of assurance that radioactive waste will be safely disposed of and that the schedule is not likely to be very different from DOE's reference schedule, the public credibility of the nay-sayers could be strongly eroded. The extent to which this result obtains, however, will depend importantly on whether the proceeding itself is perceived to have been complete and objective and whether the comissioners appear to have reached a judgment based on the evidence or on previously held beliefs.

There can be little doubt that significant progress was made during the Carter years in understanding and beginning to address the institutional problems of nuclear waste management and in identifying and proceeding to deal with remaining technical problems and uncertainties. Much has yet to be done, however. Whether the current administration and the Congress will choose to build on the successes of the past or to chart a totally different course and how they will deal with the large budgetary demands of the program remain to be seen. The progress recently achieved could certainly and all-too-easily be reversed if the federal government—either executive branch or Congress—diverges too far from the policy course that the previous administration tried to set. The relationships are still fragile and the level of mutual trust among important participants both inside and outside the government, while higher than before, is still not very high.

NOTES

1. See the fact sheet in Appendix A of this volume for a description of the various types of nuclear waste and the quantities currently and expected to be in existence in the United States.

2. For a more complete history of the high-level nuclear waste management programs of the United States, see Daniel S. Metlay, "History and Interpretation of Radioactive Waste Management in the United States," in *Essays on Issues Relevant to the Regulation of Radioactive Waste Management* (Washington, D.C.: Nuclear Regulatory Commission, Office of Nuclear Material Safety and Safeguards, 1978), NUREG-0412; and Richard G. Hewett, *Federal Policy for the Disposal of Highly Radioactive Wastes from Commercial Nuclear Power Plants: An Historical Analysis,* U.S. Department of Energy, Washington, D.C., March 9, 1978.

3. U.S. National Academy of Sciences—National Research Council, *The Disposal of Radioactive Wastes on Land, A Report of the Committee on Waste Disposal,* Publication 519, April, 1957.

4. U.S. Atomic Energy Commission, *Draft Environmental Impact Statement,* WASH-1539, September 1974.

5. U.S. Energy Research and Development Administration, Release No. 76-355 (December 2, 1976).

6. 42 *Federal Register* 34393, July 5, 1977.

7. For a more thorough discussion of the legal authorities of federal agencies with respect to high-level waste, see Emilio Jaksetic "Legal Aspects of Radioactive High-Level Waste Management," *Environmental Law* 9:347-406.

8. President Reagan has stated his intention to abolish the Department of Energy, but, at this writing has given no indication of where the nuclear waste management program would then be located.

9. 44 *Federal Register* 38690, July 2, 1979.

10. For a full discussion of the legal basis for state and federal authority in this area, see Chapter 4 in this volume. See also Donald J. Moran, "Regulating the Disposal of Nuclear Waste," *Case & Comment* (September-October 1980), pp. 21-28; and Jaksetic, "Legal Aspects."

11. 46 *Federal Register,* 5298, January 19, 1981.

12. *Pacific Legal Foundation* v. *State Energy Resources Conservation and Development Commission,* 472 F. Supp. 191 (S.D. Cal. 1979), *appeal pending.*

13. *State of Minnesota* v. *Nuclear Regulatory Commission,* 602 F.2d 412 (1979).

14. 44 *Federal Register* 61372, October 25, 1979.

15. See Chapter 4 in this volume.

16. See for example, Philip Boffey, *The Brain Bank of America* (New York: McGraw-Hill, 1975), especially chapter 5.

17. U.S. Department of Energy, Directorate of Energy Research, *Report of Task Force for Review of Nuclear Waste Management,* Draft, February 1978, DOE/ER0004/D (hereafter cited as the Deutch report).

18. The White House, Office of the White House Press Secretary, *Statement of the President on Nuclear Power Policy,* April 7, 1977.

19. U.S. Department of Energy Press Release, *DOE Announces New Spent Nuclear Fuel Policy,* October 18, 1977.

20. It was introduced by Mr. Staggers on March 1, 1979 as the Spent Nuclear Fuel Act of 1979. H1055. *Congressional Record,* March 1, 1979.

21. Deutch report.

22. Memorandum for the Secretary of State et al. from President Carter, March 13, 1978. Appendix A of Interagency Review Group on Nuclear Waste Management, *Report to the President by the Interagency Review Group on Nuclear Waste Management,* March 1979, TID-29442 (hereafter referred to as the IRG Report to the President [Revised]).

23. Interagency Review Group on Nuclear Waste Management, *Subgroup Report on Alterna-*

tive Technology Strategies for the Isolation of Nuclear Waste, Draft, October 1978, TID-28818 (Draft), Appendix A, "Isolation of Radioactive Wastes in Geologic Repositories: Status of Scientific and Technological Knowledge" (hereafter referred to as IRG, *Subgroup Report,* Appendix A).

24. IRG, *Subgroup Report.*

25. Interagency Review Group on Nuclear Waste Management, *Report to the President by the Interagency Review Group on Nuclear Waste Management,* Draft, October 1978, TID-28817 (Draft): hereafter referred to as IRG, *Report to the President* (Draft).

26. IRG, *Report to the President* (Revised).

27. 46 *Federal Register* 13971 (February 25, 1981).

28. IRG, *Report to the President* (Revised), p. 16.

29. For the IRG's discussion of this subject, see IRG *Report to the President* (Revised), p. 74. A more complete and recent discussion may be found in U.S., Office of Nuclear Waste Isolation, *The Disposal of Spent Nuclear Fuel,* Topical Report, December 1979, ONWI-59.

30. International Nuclear Fuel Cycle Evaluation (INFCE) Summary Volume, INFCE/PC/2/9, International Atomic Energy Agency, Vienna (January, 1980), p. 21. The issue has not been put to rest. A report by the U.S. General Accounting Office in June 1981 concluded that "spent fuel does indeed create problems that make its isolation more difficult."

31. U.S. Department of Energy, Office of Waste Management and U.S. Department of Interior, Geological Survey, *Earth Sciences Technical Plan for Disposal of Radioactive Waste in a Mined Repository* (April 1970), DOE TIC-11033 (draft).

32. The president's message used the date 1985, but this was subsequently revised on the basis of more detailed analysis.

33. 46 *Federal Register* 971, February 25, 1981.

34. See P. A. Witherspoon, N.G.W. Cook, J. E. Gale, "Geologic Storage of Radioactive Waste: Field Studies in Sweden," *Science* 211 (February 27, 1981): 894-99. This paper also provides a helpful discussion of the advantages of conducting site investigation at depth.

35. Jaksetic, "Legal Aspects," p. 367.

36. IRG, *Report to the President* (Revised), p. 95.

37. See for example, Rock Reiser, Hank Schilling, Randy Smith, and John Mountain, eds., *Consultation and Concurrence Workshop Proceedings. September 23-26, 1979. The Orcas Meeting, Eastsound, Washington,* January 1980, ONWI-87.

38. The White House, Office of the White House Press Secretary, *Fact Sheet. The President's Program on Radioactive Waste Management,* February 12, 1980, p. 4.

39. 45 *Federal Register* 27366, April 22, 1980.

40. 46 *Federal Register* 2556, January 9, 1981.

41. 45 *Federal Register* 65521, October 3, 1980.

42. U.S. Nuclear Regulatory Commission, *Regulation of Federal Radioactive Waste Activities,* NUREG-0527, September 1979.

43. U.S. Department of Energy, *The National Plan for Radioactive Waste Management, Working Draft 4,* (January 1981).

44. Cyrus Klingsberg and James Duquid, *Status of Technology for Isolating High-level Radioactive Wastes in Geologic Repositories,* DOE/TIC 11207 (Draft) October 1980.

45. U.S. Department of Energy, *Statement of Position of the United States Department of Energy in the Matter of Proposed Rulemaking on the Storage and Disposal of Nuclear Waste,* DOE/NE - 0007, (April 15, 1980).

46. See for example, U.S. General Accounting Office, *Federal Facilities for Storing Spent Nuclear Fuel—Are They Needed?* EMD-79-82, June 27, 1979.

47. 46 *Federal Register* 139771, February 25, 1980.

48. 46 *Federal Register* 6677, May 14, 1981.

49. See State Planning Council on Radioactive Waste Management, *Final Report to the President, August 1, 1981.*

50. See, for example, the views of the State Planning Council on this subject in its interim report,

State Planning Council on Radioactive Waste Management, *Interim Report,* (February 24, 1981), p. 11.

51. 46 *Federal Register* 9162, January 28, 1981.

52. Letter from Omer F. Brown II, Attorney, Office of the General Counsel, DOE, to Marshall E. Miller, Administrative Judge, U.S. Nuclear Regulatory Commission, March 27, 1981.

53. Memorandum to the Secretary of Energy by President Reagan, March 20, 1981.

54. Isaac J. Winograd, "Radioactive Waste Disposal in Thick Unsaturated Zones," *212 Science* 1457-1464 (June 26, 1981).

55. 46 *Federal Register,* 35280, July 8, 1981.

56. See, for example, Mason Willrich and Richard N. Lester, *Radioactive Waste Management and Regulation* (New York: Free Press, 1977).

2

A Cross-National Perspective on the Politics of Nuclear Waste

David A. Deese

Radioactive waste management poses one of the most intriguing science policy issues of our time, and several of its most challenging dimensions are inherently international in nature. Most nations with experimental or commercial nuclear programs rely heavily on imported expertise, materials, and equipment for the entire nuclear fuel cycle. Even in the major nations that export nuclear technology worldwide, the nuclear waste disposal problem has defied solution. Not only can all nations with nuclear programs learn from each other in radioactive waste management, but their programs are directly affected by events in other nations. In fact, in a number of cases spent fuel and other nuclear products from one country are transported to, processed in, and stored in other countries.

One important way to improve energy and especially nuclear policies in the United States and other countries is through systematic comparative study of national decision-making processes. By seeking variations among the sociopolitical and institutional components of the nuclear power and radioactive waste policies of different countries, it is possible to highlight means to improve national and international responses to a seemingly intractable problem, the management of wastes from military and commercial nuclear programs worldwide.

Problems in managing radioactive waste and spent fuel have frozen or delayed commercial nuclear energy programs in several countries. Austria, Canada, India, Japan, Sweden, Switzerland, the United Kingdom, West Germany, and other countries join the United States in this dilemma. The only means to progress in many countries is the establishment of prudent and efficient federal governmental programs that build on the knowledge and experience of state or provincial governments, industry, research and academic institutions, and public interest groups.

To date, no experimental or operational facility exists anywhere in the world for the final disposal of military or commercial high-level radioactive waste. The experience of the British may be instructive. In January of 1975 the buildup of spent fuel in storage facilities at Calder Hall forced the shutdown of one reactor; it has become a de facto spent-fuel storage facility. Other storage ponds in the United Kingdom are loaded beyond design specifications with spent fuel. It is clear that careful and efficient action is required to meet spent-fuel and radioactive waste management problems now, or countries will not only be unable to expand current programs for light-water reactors, but will also face the shutdown of increasing numbers of currently operating nuclear stations.[1]

This analysis focuses on the evaluation of comparative national policy formulation processes. Mapping national programs in conjunction with social, political, and administrative structure and process and comparing the similarities and differences among them has revealed six major issues in radioactive waste management: technological bias in decision making; lack of national strategies; fragmentation of governmental power structures; crippled national regulatory bodies; complex and competing relations among local, state, and federal levels of government; and increased importance of nongovernmental actors and public participation. The first two issues are overarching, encompassing the fundamental approach to policy, whereas the last four describe more specific aspects of the decision-making structures and processes. The final issue is not analyzed separately here since it is thoroughly covered in Chapter 6. It is also included in the discussions of other issues.

TECHNOLOGICAL BIAS IN DECISION MAKING

First is the set of constraints operating on the decision-making process as a result of a strong technological bias in approaching radioactive waste management; in almost all countries the problem has been misperceived as one driven solely or almost exclusively by technology. Now, after years of neglect, the significance of social, political, institutional, and economic variables has become obvious. These variables have taken precedence in the broader national and international debates regarding radioactive waste management, public understanding of the scientific process, and the apparent lack of consensus in the scientific community on related technical questions.

As yet, no government appears to have achieved a radioactive waste management (RWM) policy formulation process that carefully blends the physical sciences and engineering on one hand, and the social sciences and law on the other. France and the Soviet Union are the only nations that have successfully maintained a highly technocratic decision-making process in RWM, although the British and Japanese approaches are still more technocratic than in other industrialized nations. The processes in Canada, Switzerland, Sweden, the United States, and West Germany have shifted dramatically. Decision making in RWM is driven, at least in the short term, by complex legal, social, and political considerations.

When national nuclear energy development programs for both military and commercial purposes were in their early stages, the overwhelming emphasis was on rapid technological development, especially of power reactors. RWM and other components of the fuel cycle such as transportation were not seen as important problems and few financial resources or career and research incentives were available.

The first general manager of the U.S. Atomic Energy Commission recently confirmed what has now become an accepted explanation for the lack of a comprehensive RWM policy:

> Nobody has been much interested in the back end of the fuel cycle. The part that probably had the lowest priority had to do with the disposal of high level radioactive waste from the chemical reprocessing operation . . . The central point is that there was no real interest or profit in dealing with the back end of the fuel cycle. . . . No one appeared to understand that if the whole system did not all hang together coherently none of it might be acceptable. Indeed, this is turning out to be the case. We seem to be years away from a real solution to the back end problem.[2]

Until the mid-1970s, the few major RWM decisions were made on a more or less ad hoc basis with little attention from senior managers. In most countries, government and industry alike assumed that RWM was not technically difficult while they accumulated increasingly larger volumes of low-level, transuranic, and high-level waste materials.

Starting in the early 1970s, however, several things happened to direct attention to the waste problem. A general increase in environmental awareness occurred both within nations and internationally, especially after the Stockholm Conference on the Human Environment in 1972. There was an increase in the political power of environmental organizations and parties which was accompanied by passage of a number of national statutes and regional and global treaties covering environmental pollution problems. After decades of neglect, interest in the specific problem of solid and liquid waste disposal was also heightened.

The oil crisis of 1973 and 1974 caused governments to shift energy supply hopes onto nuclear power as one major substitute for petroleum and other fossil fuels. Yet this surge of nuclear optimism was cut short by sharply reduced economic electric power growth rates, regulatory delay, rapidly escalating costs, and opposition from environmental and other public interest groups. RWM also began to assume increasing importance in the controversy over nuclear power. Even with a dramatic decrease in orders for future nuclear plants, it was possible to mobilize greater controversy, opposition, and intervention as national nuclear power programs grew on the basis of past reactor orders and as an increasing number of people were directly affected by nuclear facility sites and associated transportation. Wide publicity of technical, political, and management problems at the Lyons, Kansas repository site in 1970 and 1971 helped escalate RWM up to the U.S. national political arena, and environmental groups in several countries began to shift their primary efforts from issues of local siting and reactor safety to problems posed by RWM.

By the mid- to late 1970s, spent-fuel storage and RWM—once an issue of little importance to governments—had become the leading obstacle to future development of nuclear power. This created strong pressure to solve RWM problems as quickly as possible. Budgets were increased, government hearings were held, and various institutional changes were made.

In 1976 and 1977, governments in Belgium, Japan, Sweden, West Germany, and the state of California attempted to legislate solutions to problems with the back end of the fuel cycle. More recently, the Netherlands, Switzerland, and several other states in the United States have followed suit. Generally, in these laws, reactor licensing has been made contingent on prior domestic or foreign arrangements for reprocessing and/or RWM requirements. The benficial effect of these regulations was to force serious budgetary and institutional action. Unfortunately, however, they also imposed a rushed and rather blind legal answer onto an increasingly complex problem. The problem demands not the rapid selection of single ad hoc solutions, but rather a carefully reasoned government and industrial program aimed at developing the best possible technical options. Whereas milestones and check points on national RWM programs are certainly necessary, crash studies and projects have proved counterproductive.

Governments in Canada, Japan, Sweden, the United States, West Germany, and perhaps even the United Kingdom appear to be tipping the balance toward basing RWM policies heavily on social, political, and legal factors. A combination of strong political pressures, state and legislative action, and problems with past technocratic approaches to RWM policy haunts federal agencies and parliaments. In some cases, their responses now focus on the most politically expedient options for technological development and siting.

In most countries, a highly technocratic policy process may no longer work; important social, legal, political, economic, and institutional factors must be acknowledged and studied in parallel with the physical scientific questions. But overreactions—in the form of policy processes driven almost solely by social and political factors—are equally counterproductive. Social scientific studies, and especially public understanding, can only accompany and build on sound physical science and engineering.

A LACK OF NATIONAL STRATEGIES FOR THE RWM PROGRAMS

Many countries are struggling with the problems of how to formulate an effective overall national strategy for managing radioactive wastes. As a now highly politicized and emotional issue in at least Austria, Belgium, Canada, Japan, Sweden, Switzerland, the United States, West Germany, and to some extent, the United Kingdom, it has captured considerable high-level management attention. Yet, four obstacles to credible policymaking resist solution. First, fundamental conflicts continue concerning the basic steps in the technical system—spent-fuel storage, transportation, reprocessing, waste solidification, test drilling, final repository siting, and monitoring. Second, problems remain in establishing a systematic process for narrowing the number of geologic media and sites to be pursued for final disposal. Third, none of the countries in this study, with the partial exception of Sweden, has been able to integrate interested and affected publics into the decision-making process.[3] And, finally, managers have been unable to develop a schedule for program and strategy milestones acceptable to government agencies, industry, public interest groups, and concerned individuals.

Defining the basic steps for the technical system in many countries has raised major questions and controversy among the physical and social science disciplines, as well as in environmental organizations and the broader publics. The lack of attention to developing the entire fuel cycle and supporting logistical systems from the beginning of nuclear power development is causing serious problems. It has forced operation and management decision for RWM to the highest levels of government and industry and has made presentation of a coherent nuclear program difficult, if not impossible. These decisions are now also being made in national parliaments, with deep involvement of the political parties in several countries outside the United States.

With or without a firm domestic reprocessing program, many countries are unable to decide not only where and how spent reactor fuel will be stored but also how to divide responsibility between government and industry. Utility operators, increasingly caught with rapidly filling water storage pools, find that the government is unable or unwilling to provide timely assistance; in most cases, the utilities cannot act on their own without prior government permission.

Spent-fuel management also frequently requires major federal government decisions because important domestic policy, foreign policy, and international political issues are involved. Japan needs U.S. approval for its only operational spent-fuel management option through the 1980s—transfer to Britain and France for storage and eventual reprocessing. Sweden has to decide how much fuel to send to France for reprocessing and where to site a large central storage facility for interim spent-fuel storage. West Germany was unable to gain approval for the large integrated fuel cycle center planned at Gorleben or to continue full use of the low- and medium-level waste disposal repository at Asse. Test drilling continues at Gorleben for studies of final disposal, and tentative approval has been gained to site two or three interim spent-fuel storage facilities. Japan, Sweden, and West Germany could not license additional nuclear plants without explicit arrangements for handling the spent fuel to be produced.

When British reprocessing of fuel from civilian Magnox power plants was suspended for ten months in 1973, due to lack of adequate tank capacity for the wastes, the Magnox fuel at Windscale and elsewhere could not be removed from cooling pools as planned and radioactivity began leaking into the water. By January 1975 one reactor plant at Calder Hall was shut down and converted to a de facto spent-fuel storage facility. Since then, utilities with Magnox stations have been forced to keep fuel on site, and their storage ponds are filling beyond designed capacity.

This situation is also aggravated by the accumulation of spent Magnox fuel from Italy and Japan and spent oxide fuel from other countries at Windscale. The future program of spent-fuel storage will have to accommodate domestic and foreign Magnox fuel, advanced gas reactor fuel, and oxide fuel from existing and planned foreign reprocessing contracts.

For some four years the United States had no clear policy on approval for overseas reprocessing of U.S.-supplied spent fuel. These approvals were granted only on a case-by-case basis. There was wide agreement on the immediate need for some U.S. capability to implement the announced nuclear nonproliferation policy of accepting the return of some foreign spent fuel. Yet the executive branch policy of accepting limited amounts of foreign spent fuel for storage in the United States cannot be implemented until the Congress passes legislation and retrieval and storage capabilities are developed. Furthermore, this policy became controversial because the Carter administration proposal for central spent-fuel storage facilities included immediate storage capacity for domestic as well as foreign fuel.[4] Some congressmen and environmentalists objected only to the timing of away-from-reactor (AFR) storage for domestic fuel, whereas others questioned the more fundamental issue of government involvement in AFR storage for domestic fuel; they preferred to have the U.S. utilities manage their own spent fuel on site until the government was able either to provide reprocessing services or to assume responsibility for final disposal.

With the possible exceptions of France and the Soviet Union, reprocessing operations and plans are also uncertain. Many technical, regulatory, economic, and policy problems surround reprocessing, both past and future in the United States. While the Reagan Administration appears receptive to commercial reprocessing, it has stated that the federal government will not finance or subsidize operation of the Allied General Nuclear Services reprocessing plant in Barnwell, South Carolina. Britain and Japan have confronted long-term reprocessing plant shutdowns, accidents and personnel radiation exposures, and important foreign policy differences with the United States. Despite the positive finding of the British government after the 100-day Windscale hearing, technical and political problems remain with the plan for large-scale reprocessing of foreign spent fuel in the planned THORP facility.

West Germany faced a very difficult set of decisions over siting the large reprocessing plant at Gorleben. After a major hearing in March and April, Prime Minister Albrecht announced on May 16, 1979 that Lower Saxony had decided against the integrated Gorleben facility. Since reprocessing or other firm arrangements are required under West Germany's nuclear law, the forced changes in federal policy cast doubt over the licenses for all nuclear power plants under construction. The resolution has been to build a smaller reprocessing facility at the current federal research facility at Karlsruhe.[5]

The French reprocessing program is the only one that still appears to be feasible as a "commercial" venture. The government-owned firm, COGEMA, had processed 100 tons of fuel from commercial reactors in France by early 1980 and had contracts covering 6,000 tons of fuel from Belgium, Japan, the Netherlands, Sweden, Switzerland, and West Germany. These commitments cover all planned capacity through 1995, including the expansion due to be completed at La Hague by 1985.

Major uncertainties continue, however, concerning the real timing and capabilities of the program. Prices charged by the French increased from $125 per kilogram in 1972 to almost $700 in 1977 and 1978. CFDT, the leading labor union in the nuclear field, shut down the La Hague plant for several months in 1976 over occupational health and safety disputes. The union still feels that working conditions are not safe at La Hague and opposes the planned expansion and all reprocessing of foreign fuel.

Due to the strike and other delays in reprocessing older fuel, there is some doubt as to whether even the existing contracts can be met in anything that approaches the formal schedule. Other questions remain concerning how and when the vitrified wastes will be returned to reprocessing customers, whether the broader public opposition to reprocessing foreign fuel will continue, how differences with the U.S. nonproliferation policies will be settled, and what restrictions will be applied on the return of plutonium produced by reprocessing.

Additional steps in the proposed basic technical system have included solidification of reprocessing wastes, preparation of spent fuel for direct disposal, transportation links, and the establishment of final disposal facilities. Until the changes in U.S. reprocessing policy under the Carter administration, the assumed process in most or all countries was solidification and vitrification of liquid high-level wastes from reprocessing. Sweden and the United States have subsequently focused more attention on packaging technologies for spent fuel, while France, Britain, Japan, and West Germany continue to count on vitrification of reprocessing wastes. France and the Soviet Union have particularly well advanced vitrification processes. Questions raised earlier, however, by the American Physical Society study in the United States about relying on the integrity of glass waste products are also being posed by European environmental organizations. Solidification proposals based on ceramics or synthetic rocks were advanced in Australia, Britain, and the United States. Proponents claimed to have "solved" the disposal problem for high-level wastes from reprocessing, since the new waste form will last for thousand to tens of thousands of years, thus dramatically reducing the requirement for assured containment by geologic disposal media. This uncertainty over waste form introduces further complexity and disagreement into the process of developing national RWM strategies.[6] Important questions also remain open in several countries concerning how best to match the waste form with a specific selected geologic medium and how to balance emphasis for containment between man-made waste forms and the natural geologic media.

The second obstacle to formulating effective national strategies for RWM is reaching agreement on a systematic process for narrowing the number of geologic media nd sites to be investigated as potential disposal facilities. Government officials in the United States face an increasingly tight set of legal and political constraints, although the choice of possible geologic media for disposal is very broad. In the past, the United States has focused almost exclusively on horizontally bedded salt as the medium for containing high-level wastes.

In France, Sweden, and the United Kingdom, on the other hand, officials are focusing almost exclusively on granite or other local media for final disposal, a selection based on the geologic reality that it is either the only viable option or one of a very few options in these countries' more limited territories. The United Kingdom is very tightly constrained on both the availability of potentially suitable geologic media and the political opposition to drilling even for lab samples from the possible useful granite in southwest Scotland. This leaves strong potential interest in waste disposal at sea. West Germany has always planned on final disposal in salt, especially in domes, since that is primarily what is available. Many experts feel that it is unlikely that Japan has any potentially suitable disposal sites for high-level wastes when all geologic, topographic, hydrologic, demographic, and political factors are considered.

Given this combination of technical and political constraints on possible final disposal sites, governments are generally reluctant to specify their criteria for media and site selection in any degree of detail. They fear that a systematic and precise set of criteria may raise difficulties with options that they want very much to keep open. But they also need more basic data from the laboratories and test sites before they can be expected to perfect the narrowing process and the selection criteria.

There are important differences among countries in the kinds of decision-making processes currently in use to deal with this problem. The United States had shifted away from exclusive reliance on horizontally bedded salt as the reference system. The past narrow focus on only one technical option for final waste disposal was to be broadened to several different sites and media after high-level executive branch attention in 1978-1980.

In early 1981, however, the U.S. Department of Energy appeared to shift its primary emphasis back to a smaller number of options, many of which are thought to be readily accessible on federal government reservations. Basalt is under investigation at the Hanford reservation in Washington, where some option must be found for troublesome high-level wastes from military reprocessing programs. Tuffs on the Nevada test site are under study, and attempts continue to use the Waste Isolation Pilot Plant (WIPP) site near Carlsbad, New Mexico, now authorized as an unlicensed repository for military transuranic waste, as an experimental high-level waste demonstration facility. (Salt domes in the gulf coast states and bedded salt in Texas and Utah are the most likely candidates for near-term sites for commercial waste disposal that are not on existing federal reservations). This approach in favoring sites more accessible to the Department of Energy may be a formula for serious failure in the fundamental objective—tangible progress toward a sound and credible final disposal program for high-level nuclear wastes.

European countries place much more emphasis on the importance of vitrification, surface storage for twenty, thirty, or even fifty years, and eventual disposal in the limited types of media available in each country. The Japanese approach is to focus on transferring all spent fuel to Europe through the 1980s, while trying to avoid any real discussion of where and how locally produced and returned high-level wastes will be contained.

The third obstacle to RWM strategy development is the lack of effective public participation programs. The long history of secrecy and technocratic decision making in military and even commercial nuclear programs provides a sharp contrast to the emerging level of public concern in the United States and, to a lesser extent, in Sweden, West Germany, and Britain.[7] New public interest and power emerged in the wake of the movement toward "open government" and the new environmental trends. More than anything else, members of the general public in at least Western Europe and the United States simply want to know more about programs and decisions that will affect their lives. This dimension even appears in France and the Soviet Union.

Some government officials and corporate managers have assumed that greater public understanding of nuclear fuel cycle issues would increase public acceptance of the associated government programs. This is not necessarily true. Public hearings, social science research, and public opinion polls conducted in several countries indicate that increased understanding, at least about issues of nuclear health and safety, increases public indecisiveness and may also increase opposition to nuclear programs. The possible implications of this relationship, which still requires more research and elaboration, make it easy to understand why many high-level bureaucrats and politicians may be reluctant to launch serious public participation programs.

Even so, the only route to credible, acceptable programs for RWM in at least Austria, Japan, Switzerland, Canada, Sweden, the United Kingdom, the United States, and West Germany appears to be through public participation. There is, however, an important difference between providing public access to information and allowing some degree of direct public control over decisions in RWM. To date, these eight countries have offered varying degrees of information concerning their RWM programs. The Windscale hearing in the United Kingdom, the Gorleben International Review in West Germany, the KBS Report and associated reviews in Sweden, and the Interagency Review Group Report in the United States have all produced new data and insights into RWM questions. Although none of these involved any direct public control over decision making, the process of public comments and hearings exerted significant influence over major governmental decisions on important aspects of RWM policy in at least Sweden, West Germany, and the United States.

French and Japanese programs have highlighted reprocessing and minimized references to final disposal plans. To date, they have generally been successful in confining public opposition concerning RWM to specific sites or groups. Although siting problems for final disposal facilities are far from being solved in France and Japan, these location-specific problems have not coalesced into an overarching issue that strongly influences national-level nuclear policymaking.

The final obstacle to national strategies for RWM lies in the difficulty of developing a widely accepted and effective schedule for the RWM program. Major national public participation programs (Sweden), technical studies and reviews (Sweden, the United States, West Germany) and public hearings (the United Kingdom and United States) all take many months, even years, to complete. Additionally, development of new legislation for RWM or compliance with existing law, such as environmental assessment requirements, can take several years.

It is increasingly difficult for these governmental functions to respond to both the needs of industry and the demands and concerns of involved publics. The lead time and stability in the investment climate required for effective industrial operations mean that decisions to go ahead are sometimes needed before governments seem capable of providing adequate services. When executive branches of governments attempt to speed up program schedules, they frequently encounter complex bureaucratic and legal obstacles. There seems to be an elusive, but extremely important line between proceeding too slowly and too rapidly with national RWM programs. Erring too far in either direction can undercut sound social and physical scientific research efforts, public credibility, and industrial effectiveness.

The level of pressure on the governmental system for finding a rapid answer to radioactive waste disposal seems to be higher in the United States than elsewhere. Technical reasons are frequently cited as a reason to hold off on immediate action, but many actors in the government are now not so concerned about technical uncertainties since there seem to be ways to "design around" problems such as temperature. The U.S. government does not give high credence to the technical arguments for delaying progress in this area. Yet the availability of a radioactive waste disposal system is clearly being held against the long-range future of nuclear energy in the United States. Some actors in Congress have attempted to enact bills that would drive the system rapidly and perhaps arbitrarily into radioactive waste management decisions along the lines of some that have been made in West Germany and Sweden, due to their local legal requirements. There may be a sense in Congress that it is essential to get on with this job; but given the lack of program coherence and leadership in the Department of Energy, the U.S. may well begin making some of the same past mistakes over again.

Basic to approaching alternatives or revised paths to radioactive waste and spent fuel management is much greater clarity and simplicity in defining the issues, the actors, and the relevant time schedules. One clarifying step is to define three principal issues, actors, and time periods as, for example, in table 2.1.

Table 2.1   Strategic Planning for Radioactive Waste Management

Issue	Principal Actor	Timing of the Most Critical Decisions
Low-level wastes and mill tailings (including military)	States	Immediate (1981-83) Planning and negotiations for new regional disposal sites
Spent fuel management	Utilities	Near-term spent fuel storage
Research, demonstration and siting of military and commercial high-level waste disposal facilities	Federal government	Immediate—Planning and exploration
		Near-term—Intergovernmental coordination, budget priorities, and field tests
		Longer-term—Siting and construction

This is one way to frame discussion of the means and substance of effective strategies for radioactive waste and spent-fuel management. Specific questions surrounding each of the three central issues can be listed and discussed, including a useful division of the questions into two categories: those related to more technical aspects of physical and social sciences, and those of ethics and social responsibility.

Given these two sets of questions and discussions for each issue, the focus can shift to related institutional mechanisms. If the states continue to lead in low-level waste management, what role should the federal government have in case new sites are not expeditiously developed? What role should the Nuclear Regulatory Commission (NRC) have in state and federal sites for low-level waste? What role should New Mexico and the NRC have in the WIPP?

If the utilities assume primary responsibility for spent fuel management, while awaiting results of federal research, development, and demonstration (RD&D) on high-level waste disposal, what action is required by the federal government (especially Congress) to establish a reasonably secure and consistent operating environment for the utilities, including maximum possible confirmation of how and when responsibility for final disposal will be assumed by the federal government? (The recent past case of Tennessee Valley Authority, which dropped plans for AFR storage, is instructive.) What action should the federal government, especially Congress, take to execute an apparent consensus on the return of foreign spent fuel as appropriate to gain important nonproliferation advantages?

If the federal government is effectively to manage the long-term problem of RD&D, final siting, and full-scale operation of a high-level radioactive waste disposal system, what new interagency and intergovernmental coordinating mechanisms are required? Assuming that some provision is necessary for final settlement of disputes in the consultation process with the states, when should the decision go to Congress or the president? What links are appropriate between : (a) well-established milestones in the RD&D and final siting program in high-level waste and (b) the continuation of interim spent-fuel management programs; reactor construction, licensing, and operation; the overall size of a U.S. nuclear energy program; and possibly the general allocation of government R&D budgets in energy?

Whenever possible, actions and responsibilities of institutional actors should be assigned after a careful assessment of required capabilities and to a specific institutional time period in order to assure coordinated functions. Useful criteria for guiding decisions and recommendations on institutional capabilities include:

- Being functionally specific

- Assuring continuity in financing

- Assuring strong management and staff (carefully considering the Kemeny report on Three Mile Island)

- Having the capacity to call on outside expertise

- Avoiding excessive delay in replacing or establishing institutions

- Being explicitly designed for planning, R&D or implementation

- Having a balance in emphasis between physical and social sciences

- Being adaptable to changing conditions, but with continuity in personnel over time

- Having access to broader resources and views of the issue (financial trade-offs; foreign policy; licensing authority)

- Making use of optimal public/private organizational blend (specification of how and when outside contracts are to be used)

FRAGMENTATION OF GOVERNMENTAL POWER STRUCTURES

The third major issue is the severe fragmentation of power and authority within and among federal government structures. As the perceived importance of the back end of the nuclear fuel cycle increased rapidly over the past few years in Austria, Britain, France, Japan, Sweden, Switzerland, the United States, West Germany, and even the Soviet Union and some less-developed countries, new governmental institutions were created and existing ones acted to capture a role in solving the problem. This fragmented power base has been at least partially responsible for the lack of effective executive and legislative leadership in countries with major nuclear power programs, except France and the Soviet Union.

There are, however, important cross-national differences both in the degree of fragmentation within the executive branches and the degree to which power has been diffused beyond the executive branch in this issue. These differences must be analyzed in the broader context of differences in basic political and private structures and public policy-making processes. Even within Western Europe, but especially between the United States or Japan and Western Europe, for example, there are different degrees of separation between executive branch agencies and parliamentary structures. The complete separation of membership between executive and parliamentary structures found in the United States and Japan does not exist in the United Kingdom and West Germany, where an important member of a federal ministry may also be in Parliament or the Bundestag.

The balance of public as opposed to private ownership and control over the nuclear fuel cycle varies widely among countries. Ownership patterns are sketched in table 2.2 for the nuclear fuel cycle and electric utility industries. There seems to be a direct correlation between the level of government owner-ship and the health of nations' nuclear industries. This correlation also extends to the salience of RWM and other issues associated with the nuclear fuel cycle.

In the Soviet Union and France, where the states own most components of the fuel cycle, the RWM issue is less salient and the nuclear industry is in reasonably good condition. In the countries with dominant, or very influential, private sector interests, such as the United States, West Germany, and Sweden, RW and spent-fuel management are highly conspicuous issues, and at least parts of the nuclear industry face grim prospects. Varying mixes of public and private arrangements in Canada, Japan, and the United Kingdom are asso-ciated with limited visibility of RW and spent-fuel management issues and some growth prospects for the nuclear industry.[8]

We have encountered a proliferation of institutional changes in executive branch bureaucracies in many countries with parallel activity in the parliamen-tary structures and the national-level political parties of some countries. Although the involvement of court systems (even in France) in issues related to RWM is increasing, this has been influential in national policy decision to date only in the United States and West Germany.

Table 2.2   Patterns of Ownership in the Nuclear and Electric Utility Industries

	USA	Japan	FR Germany	Sweden	Canada	France	UK	USSR
Enrichment	S	m	m	—	—	S	S	S
Fuel fabrication	P	P	m	m	S	m	S	S
Electric utilities	m*	m*	m	m*	(S)	S	S	S
Architect engineering	P	P	P	m	S	S	S	S
Reactor supply	P	P	P	m	S	m	m	S
Reprocessing	P	P	m	—	—	S	S	S
Waste disposal:								
R&D	S	S	m	m	S	S	S	S
Implementation	m	m	m	?	S	S	S	S

Key S = state (central government ownership)
     (S) = state (provincial government) ownership
     m = mixed private and state (central and/or provincial) ownership
     m* = mixed ownership, in the sense that some institutions are privately, others publicly owned, e.g. in the USA, 77% of electricity is generated by private utilities, 23% by public utilities
     P = private ownership
     ? = undecided
     — = not relevant

Source: Adapted from Mans Lönnroth and William Walker, "The Viability of the Civil Nuclear Industry," a report for the International Consultant Group on Nuclear Energy, 1980.

The interaction of these national-level executive, parliamentary, party, and court structures with national regulatory bodies, local and state governments and political parties, nongovernmental actors or interest groups, and industrial bodies will be considered below.

The splintering of power among executive branch agencies in the United States and West Germany, and to a lesser extent in Japan, Sweden, and the United Kingdom is contrasted with the still relatively unitary situation in France and the Soviet Union.

The United States

Organizational complexity is a serious problem in the United States: independent regulatory agencies are important actors; several executive branch agencies play important roles; various congressional committees have important powers; and in contrast to France, the court system is a serious route to challenging the government program at almost each step.

The degree of fragmentation within and beyond the U.S. executive branch in RWM decision making is greater than in any other country. This is especially the case in the executive departments and agencies and in the congressional committees. A dramatic shift has occurred since the days of highly centralized and secretive operations of the Atomic Energy Commission and the Joint Committee on Atomic Energy. The once symbiotic relationship between these two bodies has evolved into a situation where as many as a dozen separate actors in the executive and congressional branches of government have significant influence over decision making in this area. The level of competition within each branch is exceeded only by that between the two.

U.S. decision making in RWM has been in a state of flux over the past four years. As the central line actor in this area after the demise of the AEC, the Energy Research and Development Administration conducted early RWM policy reviews in 1976 and 1977. Mid-level managers could see the important implications for domestic and international security, energy, and environmental protection issues. Internal studies by the successor agency, the Department of Energy (DOE) in 1977-1978 led to the formulation of the president's Interagency Review Group (IRG) study in 1978-1979. Under DOE leadership, the IRG succeeded in greatly broadening the decision-making base to involve not only thirteen executive branch departments and agencies, but also participation by individuals in Congress, state governments, public groups, universities, and research institutions. Now, however, the Reagan administration places much less importance on DOE. It is highly likely, however, to find that it cannot strengthen the nuclear power without progress in RWM.

As the number of involved perspectives and organizations increased in the late 1970s, there was not a proportional growth of DOE and White House leadership on the issue. DOE managed to hold on to the central line responsibility for RWM against the challenges of either an independent government authority or coordinated, independent programs in several agencies. Even so, it faced internal management challenges (such as the complex management structure of field operations and regional offices) and equally difficult external obstacles in the diffusion of decision-making power in RWM to other agencies (such as the Bureau of Land Management in the Department of the Interior) and to congressional committees, the lack of strong presidential support, and the difficult task in winning direct public credibility and support.

Beyond an effective working relationship with regulatory agencies, which will be discussed below, DOE needs at least three things to be effective in implementing a RWM program. It must receive consistent and direct support from the president, adequate budgets, and some legislation from Congress to have the authority and resources to lead the national RWM policy. The probabilities of this happening in the Carter administration appeared to decrease in proportion to the time taken in getting the IRG recommendations into presidential memorandum form, through the White House offices to the president, and finally into a presidential announcement of U.S. RWM policy in February 1980.

Power in RWM policymaking is also dispersed among various congressional committees, and most legislation proposed to date is very parochial in its approach. Table 2.3 maps the very complex structure of congressional committees that influenced nuclear policy in the ninety-sixth Congress.

West Germany

The West Germany system is federal preemption with the states acting on behalf of the federal government. All elements of the back end of the nuclear fuel cycle are handled by the industry, with the exception of disposal which, under the fourth amendment of 1976 to the Atomic Energy Law, remains the responsibility of the federal government. The government has tasked a national laboratory, PTB, with responsibility for building and running final disposal facilities.[9] This is similar to the organizational structure in Japan.

Work has been started on drilling and testing at the Gorleben facility, which includes assessment of the possibility of disposing of spent fuel. The most recent status of the project is abandonment of the facility as a comprehensive site for all aspects of the back end of the fuel cycle, with a compromise that a licensing proceeding will continue for using the Gorleben facility as a drilling and test site and possibly a future storage and disposal site.

The agreement reached by the federal government and the states will apparently allow reactor licensing to continue since there is now a prospect for progress in waste management, although it is clear that interim spent-fuel storage is required. The states of Lower Saxony and North-Rhine Westphalia have agreed that they may provide sites for spent-fuel storage while densified storage will be installed in most nuclear power plants around the country.

Licensing is done by the states for the German Federal Government under the supervision of the Federal Ministry of the Interior. The state authority calls on outside experts for safety reports; the Federal Ministry relies on two permanent advisory commissions: the Reactor Safety Commission and the Radiological Protection Commission. The PTB is responsible for licensing interim spent-fuel storage facilities. The DWK, the German Society for the Reprocessing of Nuclear Fuels, a private company composed of 12 utilities, applied for such a license for Gorleben. DWK has also applied for a license from the PTB for a spent-fuel storage facility at Ahaus. This would be built in cooperation with an engineering firm and would also require a construction license from the state.

In 1976 guidelines were given for interpreting new atomic energy amendments, with the rule that nuclear stations can be licensed only if companies have produced sufficient evidence that spent fuel can be managed safely. Not later than the granting of the first construction license, a German utility operator must show arrangements for the first six years of production of spent fuel. This also applies to plants with construction or operating permits.

The federal government now plans the back end of the fuel cycle in five steps. The first is storage on a long-term basis at reactors. Next is a thirty-year period of storage at a site such as Gorleben. Third is a version of longer-term storage for up to fifty years using some form of dry or containerized storage facility, providing time for the final steps. Next is reprocessing on a large or small scale, with ten or fifteen years of research and development on the disposal operations for spent fuel or reprocessed wastes. Finally, disposal is to be done on a permanent basis with or without reprocessing.

The total number of usable salt domes on geographic and geologic ground in Germany was at one time set at twenty-six. Based on a number of criteria, a shorter list of possible salt domes for disposal sites was constructed, but this list ignored political criteria. For example, there never was any intention on this basis to investigate the salt dome at Gorleben; this occurred only because the site was forced by the state government of Lower Saxony. The comprehensive approach to including all steps in one site was also forced on the private sector by the government. The International Gorleben Review was essentially aimed at a specific site and concept which had been previously rejected by the group of four private companies that were involved directly in this operation.

By 1980, there had been only eight hydrologic drillings by PTB and twenty-five foundation drillings into the soil by DWK at the Gorleben site. Little experimental work had been done on assessing the longer-term risk of escape by radionucleides from the underground facility. Tremendous pressure was thus placed on the West German industry to act. This was heightened by a series of court decisions which almost reached the point of directly linking reactor operations with waste management progress.

Table 2.3 Committees in the 96th Congress That Set Nuclear Policy

Committee	Subcommittee	Jurisdiction	Chairman
Senate			
Commerce, Science & Transportation	Science, Technology and Space	Transportation of nuclear materials	Howard W. Cannon, (D.-Nev.)
Energy and Natural Resources		Energy policy & R&D	Henry M. Jackson (D-Wash.)
	Energy Research & Development	Nuclear & nonnuclear energy R&D and Energy Department nuclear policy	Frank Church (D-Idaho)
	Energy Regulation	Nuclear insurance	J. Bennett Johnston Jr. (D-La.)
Governmental Affairs		Federal governmental organization	Abraham Ribicoff (D-Conn.)
	Energy, Nuclear Proliferation, and Federal Services	Nuclear proliferation	John Glenn (D-Ohio)
Environment & Public Works		Nonmilitary environmental regulations & control of nuclear energy	Jennings Randolph (D-W.Va.)
	Nuclear Regulation	NRC oversight	Gary Hart (D-Colo.)
Appropriations		Appropriation of the revenue for the support of the government	Warren G. Manuson (D-Wash.)
Foreign Relations		Arms control, oceans, & the international environment; nuclear export policy	Frank Church (D-Idaho)
Interstate & Foreign Commerce		Regulation of nuclear facilities & use of nuclear energy	Harley O. Staggers (D-W.Va.)
House of Representatives			
Interstate & Foreign Commerce		Regulation of nuclear facilities & use of nuclear energy	Harley O. Staggers (D-W.Va.)
	Energy & Power	Regulation of nuclear facilities & special oversight functions with respect to all laws, programs, & gov't. activities affecting nuclear energy	John D. Dingell (D-Mich.)

Table 2.3 Committees in the 96th Congress That Set Nuclear Policy *(Continued)*

Committee	Subcommittee	Jurisdiction	Chairman
Interior & Insular Affairs		Regulation of domestic nuclear energy industry, including R&D, reactors, & nuclear regulatory research	Morris K. Udall (D-Ariz.)
	Energy & the Environment	Special oversight functions with respect to nonmilitary nuclear energy & R&D, including disposal of nuclear wastes	Morris K. Udall (D-Ariz.)
Science & Technology		All energy R&D, special oversight for all nonmilitary R&D	Don Fuqua (D-Fla.)
	Energy Research & Production	Legislation & other matters relating to R&D & demonstration involving nuclear fission, fusion, electric energy systems & nuclear physics	Mike McCormack (D-Wash.)
Government Operations		Budget & accounting measures other than appropriations; efficiency of governmental operations; Intergovernmental relationships between U.S. & states & municipalities	Jack Brooks (D-Tex.)
	Environment, Energy & Natural Resources	Oversight & organization of Energy Dept. & Nuclear Reg. Commission	Anthony Toby Moffett (D-Conn.)
Appropriations		Appropriation of the revenue for the support of the government	Jamie L. Whitten (D-Miss.)
	Energy & Water Development	NRC appropriations	Tom Bevill (D-Ala.)
Merchant Marine & Fisheries	Fisheries & Wildlife Conservation & the Environment	National environmental policy	John B. Breaux (D-La.)
Armed Services		Military facilities	John Stennis (D-Miss.)

Japan

Japanese policy in RW and spent fuel management is vague and unformulated. In 1976 the Atomic Energy Commission was split, with nuclear regulation and the development of new technology assigned to the Atomic Energy Commission and other aspects controlled by other bodies. The Advisory Committee on RWM Technology was established to review technical activities.

Following reorganization in 1976, the nuclear industry created a RWM center for R&D and disposal operations, involving low-level wastes on land and at sea. The center is designed to integrate government and industrial work, conduct tests of low-level waste dumping at sea, and perform final disposal operations under contract to the government or industry. Their efforts specifically include a detailed feasibility study of social, legal, economic, and technical components of RW disposal.

The background for this problem is the precarious general energy situation in Japan. Earlier predictions for nuclear power capacities as of 1985 were 33 gigawatts (Gw), which will certainly not be reached under current conditions. Japan is most likely to have 20 to 25 Gw by 1985. The predicted levels of oil and liquified gas supply are also likely to be well below past estimates of requirements. Disruptions in the world oil market in 1973-1974 and in 1979 caused urgent concern over energy security. The Japanese public, therefore, has mixed feelings even after the Three Mile Island accident. It may also be important that English is not widely spoken and recognized, as in France, so that some of the events and attitudes of other countries are not transmitted immediately to the Japanese public. It is the case, however, that radioactive waste management is becoming a critical factor in the Japanese nuclear power program. This is especially true in two areas: the utilities, in sending spent fuel to Britain or France for reprocessing, must specify to the Japanese government how return wastes will be handled; and the second reprocessing plant which is due to be constructed in 1990 has to be accompanied by a management plan for expected high-level wastes. In both cases, the government has avoided discussing radioactive waste management implications, but they now realize that this policy cannot be continued.

With respect to radioactive waste disposal at sea, Japan is in the process of ratifying the London Convention of 1972. The program for disposal of low-level radioactive waste at sea parallels European plans, but sharp political opposition continues to arise from territories in the Pacific.

The Science and Technology Agency is in the position of giving overall guidance to the Power Reactor and Nuclear Fuel Development Corporation (PNC), which does nuclear fuel cycle development activity, and the Japan Atomic Energy Research Institute (JAERI) on the basic research program for high-level waste management. The PNC is studying the subseabed disposal of RW. Under the powerful Ministry for International Trade and Industry is the

Government Research Institute in Osaka, which is tasked with implementation. The only formal government policy statement in this area, made in 1972, said that the treatment of high-level waste is the responsibility of the institution which does the reprocessing, based on technologies demonstrated by the PNC and JAERI. The technological verification of this waste management process and the actual disposal activities will be conducted by the government. The current timetable is for a decision on a disposal methodology in the early 1980s, with demonstration tests by 1985. The emphasis now is on establishing concrete responsibilities for institutions to replace the current confusing sharing arrangement between several different bodies.

Much of the discussion of the back end of the nuclear fuel cycle in Japan has been driven by proliferation policies. There are questions of local opposition in both Japan and Europe to the shipping of Japanese spent fuel for reprocessing in Europe; of U.S. permission and the implications of refusal for the shipment of U.S.-origin spent fuel from Japan for reprocessing in Europe; and of the return and management of plutonium from these reprocessing operations in Britain and France. Japan accepts the international control of plutonium if there are reasonable criteria for release as needed for research and development activities. In spent-fuel storage, the utilities' storage ponds are filling up rapidly, and while the public does not oppose storage of the fuel per se, there is opposition to the increasing quantities.

The government has given permission for private entities to join in reprocessing activities. Chemical companies and the utilities formed a company to build the second reprocessing plant. Government responsibility in waste management continues to be firm for the final implementation stages, but the earlier research and development stages will continue to be done in private research institutions.

The central storage of spent fuel has not yet been raised as a central policy item, but it is possible that the second reprocessing plant will be started with a storage facility in the same site. The utilities in Japan are interested in the concept of Pacific Basin storage of spent fuel, but the Japanese government is only willing to participate in serious joint studies of this concept if it does not appear to the public to take the pressure off the need for a site for the second reprocessing plant. There may well be a demand for some centralized spent-fuel storage system as a backup facility for countries like Japan and West Germany which may be unable to get public acceptance for siting a facility in their own country. Siting is a very serious problem for the second reprocessing plant, but a few remote areas are quietly inviting such a facility. Reprocessing and waste disposal sites will only be accepted if the local public sees them as part of the broader national interest. The impression that one receives from this overall situation is a very ad hoc and informal process in a problem area that demands a much more formal and tight structure.

Sweden

The distribution of authority over the back end of the fuel cycle in Sweden is only partly settled. It is, however, constrained by the overall characteristics of Swedish administrative structure which include the following:

- The utility industry is dominated by the Swedish State Power Board, SSPB (50% of production capacity). Two other groups also operate nuclear power reactors: the municipally owned Sydkraft and the privately owned OKG.

- The nuclear fuel supply company (SKBF) is a wholly owned subsidiary of the utilities. The chairman of the board is, however, appointed by the government.

- Swedish agencies, including the SSPB and regulatory agencies such as the Nuclear Safety Inspectorate (SKI), are very autonomous. A minister cannot formally give a direct order to an agency.

- The administration is a relatively closed system, offering few openings for intervenors. A court cannot intervene in administrative decisions. Information is, however, generally made available to the public.

- Decisions under acts such as the Stipulation Act are taken by the government.

- The national government has the final say in the establishment and siting of nuclear facilities.

- Local authorities have the right to veto all land-use proposals.

- Test drillings (for geological data) require the approval of the landowner but not of the local authority. The Swedish state is the owner of large amounts of land through the National Forest Agency.

The overall administrative structure is outlined in table 2.4. The main actors in Sweden are the utilities, SKBF, the national government, the local authorities, and the SKI. The division of responsibility between the utilities and the state has not yet been determined. This issue is currently under study by a commission. Also, the French and the U.S. governments have, through various decisions over reprocessing, transportation, and other elements of the fuel cycle, a major influence on the back end of the Swedish fuel cycle.

It is probably fair to say that, compared to other countries, the authority in the Swedish administrative structure is centralized enough to implement the various facilities for the back end once the relevant design characteristics have been worked out. There are essentially no risks of the type of stalemates between various authorities that have plagued the United States and to some extent the West German programs. The only real complication could come from local authorities, but the national parliament can always, at least in principle, enact a special law which would give the central government preemption over siting and the right of veto of local authorities. Such preemptions already exist for other facilities such as roads, railways, and government buildings.

The possibility that such a law might be necessary should not be excluded. Test drillings as part of an intensive geological survey of possible sites have started at various places in Sweden. The test drillings only require the approval of landowners, and have so far mainly taken place on state-owned (or utility-owned) land. Several local authorities have, however, declared that they do not want to have the final disposal plants located within their territory. One of the problems of the siting decisions is that it touches at the heart of what is argued to be the weakest link in the safety system, namely, the ability to find a site with acceptable characteristics and to predict the geological conditions.

Table 2.4   Institutional Structure of the Back-end of the Swedish Nuclear Fuel Cycle

Facility/Activity	Owner/Operator	Licensed Under	Regulatory Agency
Reactor ponds	Utility	AEL	SKI
		SSL	SSI
CLAB (= AFR)	SKBF	AEL	SKI
		SSL	SSI
		ML	SNV
		BL	
		VL	
Transport from ponds to CLAB	Special shipping company	SSL, AEL	SKI
Transportation CLAB-abroad		As above	
Reprocessing	Cogema	French legislation	
Vitrification	Cogema	French legislation	
Uranium	Utility	As fissile material	
Plutonium return	Utility	Not yet completely specified	
LLW, LLW return	Utility	SSL	SSI
Storage of vitrified waste	Not yet specified	SSL	SSI
Encapsulation	"	"	"
Final repository	State		

KEY TO ABBREVIATIONS:
SSL = Radiation Protection Act   ML   = Environmental Protection Act
SSI = Radiation Protection Agency   SNV = Environmental Protection Agency
AEL = Atomic Energy Act   CLAB = AFR-storage
VL   = Water Act
SKI = Nuclear Safety Inspectorate

COMMENTS:

Reactor ponds—Can be expanded without problems.

CLAB—Local veto on siting. All necessary licensing decisions have been taken, the government has granted construction permits at Simpevarp. AFR storage is judged less expensive and safer than expanded ponds.

Transport from ponds to CLAB—Transport will be made by special ships, designed by SKBF, between reactor sites and AFR. The ship will be mainly owned by the utilities, but, for French licensing reasons, may also include a French role.

Transportation CLAB-abroad—Permission by Swedish government needed for transportation out of Sweden. Permission by U.S. government needed for reprocessing (MB-10s) of irradiated U.S. enriched (or origin) fuel. Operator of ship (see above) not yet clear.

Reprocessing—Contract includes *force majeure* clause. If UP3-A delayed then spent fuel will be stored at La Hague. Exchange of letters French-Swedish governments that streams (waste etc.) out of La Hague will be accepted back into Sweden has occurred.

Vitrification—Agreement Cogema-SKBF on glass characteristics and contents necessary but not yet signed. If Cogema and SKBF cannot agree on specifications reprocessing will not occur and spent fuel will be returned.

Uranium—Returned to Sweden.

Plutonium return—Swedish ownership implies Swedish control of use. French veto right on use in Sweden. Plutonium stored at La Hague. Fabrication of Mox fuel not settled (INFCE etc.).

Storage of vitrified waste—Storage for 30 years of vitrified high level waste according to proposal. Ownership could be SKBF or the state. Local authority approval required.

Encapsulation—Local authority approval required for construction of utility.

Final repository—National government has overall responsibility. Local authority approval required for construction.

SSL—Covers all radioactive materials and activities.

AEL—Covers fissile materials/activities.

Source: Adapted from: Lönnroth and Walker, *The Viability of the Civil Nuclear Industry.*

The United Kingdom

There are some parallels between reprocessing and other aspects of the nuclear fuel cycle in Britain with those in France. The relationship between industry and government, however, is very different in both the constitution and institutional arrangements. Britain also lacks the clear separation between the legislative and executive branches of government that are found in the United States. In Britain, the members of Cabinet are also members of Parliament and there is no equivalent of the U.S. administrative structure. The prime minister has less executive power than the U.S. president, and the departments are much more important. Although it is governed as a whole, Britain is not administered as such. Wales and Scotland have their own ministers, and the split responsiblity produces considerable delay in attempting to coordinate a decentralized system. Finally, local and regional powers are not as important as in the United States or West Germany.

The British government program on radioactive waste management has been described in two white papers. The most recent, in December 1978, structured the administrative responsibilities and programmatic elements, with considerable tightening of the system over what existed previously.

The central Department of the Environment maintains the strongest hold over this decentralized system. Formal responsibilities in this area are divided between the Department of Evironment; the Ministry of Agriculture, Fisheries, and Food; the Department of Energy; and special responsibilities of the Health and Safety Committee of the Department of Employment. This produces a gray area of general environmental responsibilities in the government (see fig. 2.1). Trade union representation is particularly strong on the Advisory Committee on Safety of Nuclear Installations.

Policy	Advisory	Executive
Department of Environment	Radioactive Waste Management Committee	Atomic Energy Authority
Ministry of Agriculture, Fisheries and Food	National Radiological Protection Board	British Nuclear Fuels, Ltd.
Health and Safety Committee	Advisory Committee on Safety of Nuclear Installations	

Figure 2.1 Radioactive waste management responsibilities in the British government.

This structure also has responsibility for regulating radioactive waste disposal at sea.

The British situation poses many historical parallels with that in France. The nuclear programs both had a military beginning; decision making was centralized at a very high level; the level of secrecy was very high; and the programs were designed and conducted very much as technological endeavors. The United States situation is somewhat similar but further along on the same pathway; West Germany and Sweden have been much more affected by events in the United States than has the United Kingdom. Technological developments in these programs were initially aimed at the production of plutonium for military uses. In some aspects, the movement to civilian applications of these technologies has never been completed. In Britain, for example, liquid wastes from military reprocessing operations are still mixed with those from civil reprocessing. This poses national security obstacles to both structuring administrative responsibilities in RWM and acquiring the data about civilian applications.

France

A combination of personal relationships and political leadership, administrative and legal structure, and political coalitions seem to explain why France has been able to maintain a centralized and active nuclear program.[10] The coalition supporting Valery Giscard d'Estaing won the 1978 National Assembly election. In doing so, it left behind a left-wing coalition and particularly a Socialist Party in greater disarray than it had been for years. By mid-1978, Valery Giscard d'Estaing was the leader of one of the strongest governments among the Western democracies. He was also personally committed to the success of the French nuclear development program. At a government energy policy meeting in the summer of 1978, one of the participants noted the American stalemate and asked a colleague why the same thing had not happened in France. "That's simple," was the response, "We have the President; we have the *preféts;* and we have EDF." The new administration of Francois Mitterand, however, is introducing important changes, especially in the role and powers of the *preféts.*

The French nuclear program did, indeed, have the President Giscard d'Estaing. But on a day-to-day basis, it was at least as important that the program's chief architect and guiding spirit—Andre Giraud, former head of the Atomic Energy Comission—was the head of the Ministry of Industry, which has ultimate line responsibility for the program's execution. Both the Atomic Energy Commission and the EDF report to the Minister of Industry. Giraud was also seen as one of a handful of contemporaries and former peers for whom President Giscard d'Estaing had deep respect, especially for his technical competence.

By way of comparison, in the United States one has to go back to President Eisenhower's first term and Lewis Strauss' simultaneous positions as head of the Atomic Energy Commission and advisor to the president to find a parallel. For a few years in the 1950s, Strauss controlled atomic energy development in the United States in what seems to have been much the way that Andre Giraud did, for a time being, control it in France.

But there is more behind the momentum of the French nuclear program than personalities and personal relationships. There is also institutional structure. Enormous operational power has been concentrated in two complementary agencies: CEA and EDF; in the former for the development of fuel cycle facilities and for progress on breeder reactors; in the latter for LWR construction and operation. Once again, the years of AEC-JCAE sovereignty in the United States are suggestive.

The key point, though, is that in France the nuclear development program has been and is likely to remain a high-priority objective of a strong, centralized government. On top of this, the administrators responsible for its success have an organizational apparatus shaped for concentrated fiscal and technical resource deployment and, above all, for decisive action.

These administrators also "had the *préféts.*" In the regions outside Paris, the *préféts* were the representatives of the central government with extensive power and authority for maintaining law and order. Most of the time, the *préfets* were able to use this power by deploying local law enforcement authorities to block the activities of antinuclear demonstrators at the most critical sites. The future of this system, however, remains in doubt under the new administration of Mitterand. At a minimum, the *préféts* will be replaced by elected officials in a more decentralized administrative structure. The actions of the *préféts,* and hence the government policy objectives these actions support, were also buttressed by a further reason for the relatively trouble-free progress of the French nuclear program—the French judicial machinery. In stark contrast with the United States and West Germany, the French judicial machinery has been generally hostile or indifferent to objections to government nuclear projects. The attitude of the French courts on nuclear power plant construction has been consistent with a long-standing bias toward the side of the administration when asked to make trade-offs between public and private interests relating to officially sponsored projects. In such instances, French courts nearly always come down on the side of what they consider to be the public interest, namely, the position of the administration. Consequently, French antinuclear forces, cast in the role of promoting "private" interests, have been almost uniformly unable to find the de facto allies on the bench that their American and West German counterparts appparently continue to discover. An important example was the May 1979 rejection by the Conseil d'Etat of all eighty-nine petitions made by various ecological and antinuclear groups against construction of Super-Phenix.

THE ROLE OF REGULATORY AGENCIES

In 1979 the U.S. Nuclear Regulatory Comission (NRC) decided to conduct a "confidence" proceeding to determine whether nuclear energy should go forward based on the adequacy of the current Department of Energy program for waste management, and if it were to go forward, with what steps. This is an area where it would seem to be constructive to link nuclear power programs directly with interim waste management, but weakly or not at all with final disposal.

The U.S. regulatory agencies are charged with providing criteria and guidelines for the Department of Energy (DOE) program. One crucial question, in the United States and elsewhere, is how many different sites and types of geologic media have to be submitted by the DOE to the NRC for investigation and licensing.[11] The DOE also must know how much "in situ" data is necessary before a license for a disposal facility will be issued. The NRC must have continued access to DOE data and the internal technical capability to assess the DOE programs to prevent actions by DOE that could preempt later NRC decisions. And, as highlighted by the IRG, the Environmental Protection Agency (EPA) must accelerate its schedule for general criteria and standards development if it is to carry out its mandated functions in spent fuel and RWM.

In the Federal Republic of Germany there is a clear separation of powers between the legislative, administrative, and judicial branches. Since 1973 there has been a clear division in research and development between the Ministry of Research and Technology, the Ministry of Economics, and the Department of Interior that does the licensing through delegation to the states. The Department of Interior has ultimate responsibility and has established several commissions that give expert advice; the relevant minister of state actually gives the license in the name of the federal government.

In Britain, the Department of the Environment has regulatory authority. The responsibilities for verification of new disposal sites go to the different branches of the Department of Environment, including that in Scotland. Although in dispute, the final word on the use of a site should come from the central department. Licensing a site and determining proof of safety are the responsibilities of the Department of Employment and the Nuclear Inspectorate. Criteria are to be developed by the Radioactive Waste Management Advisory Committee of the Department of the Environment, but these are advisory only in the form of published reports and public hearings.

In France the licensing power is concentrated at the top of the political structure. Until 1975 the same administrative body did both promotion and regulation. The Ministry of Industry still has the ultimate responsibility over both aspects of the nuclear power program if disagreement arises. The Council of State gives final approval on the licensing process. Despite the formal independence of these different institutions, there is strong pressure to support the programs of the central government.

Regulatory effectiveness is one example of an area where RWM is enmeshed in more fundamental challenges to the current governmental and administrative systems in several countries. This applies in at least West Germany, Britain, Japan, and Sweden, as well as in the United States. The 1970s became a decade of dramatic environmental awareness and action at a pace which was unmatched by governmental response.

RWM typifies, in at least some aspects, the class of energy and toxic wastes problems that is severely stressing the capacities of current regulatory agencies. Public expectations seemed to far exceed regulatory capabilities. These problems are relatively new and large scale when compared with the levels of expertise, staff, authority, budget, and management structure of agencies such as the U.S. NRC. Similar results were found in other countries. Sweden, for example, passed the Stipulation Act in 1976, which required the reactor operator to show how and where spent fuel or RW could be "finally stored with absolute safety" before any new operating licenses could be issued.

Six specific problems have been encountered frequently by regulatory agencies. First is the scheduling and phasing of regulatory programs under the pressure of operational demands. The regulatory agencies have attempted to catch up to and lead—rather than respond to—developmental programs. This has been driven by events in the political system in at least Japan, Sweden, West Germany, and the United States. Direct links between reactor licensing and large steps in RW and spent-fuel management in the first three countries, and indirect links in the United States, have placed tremendous pressure on regulatory systems. The number of unresolved technical, social, political, and institutional questions of RWM argue for a different approach—a series of carefully coordinated programmatic steps with several deadlines which are enforced by regulatory agencies, over approximately five to fifteen years.

This task is particularly difficult where several ministries and commissions have large areas of regulatory responsibility. Political leadership is required to ensure that regulatory requirements by different agencies are completed on time (leading rather than following development programs) and are complementary and consistent. This includes ensuring that public hearings are conducted on schedule and on the relevant issues and sites. The Gorleben International Review provides a crucial lesson: where personal politics or the political process dominate, effective public reviews of technical programs will be difficult, if not impossible. The entire review process in West Germany was conducted on a concept and a site that were forced by the government on the four companies conducting the program.

The second regulatory problem is criteria development. Progress in RWM programs has been extremely difficult when regulatory bodies have not provided criteria in advance. This is especially the case for drilling and other site characterization programs. Beyond the opposition encountered in Canada, Japan, Sweden, Switzerland, West Germany, and the United States, even the United Kingdom has had severe problems of access to sites for even very early test drillings because no guidelines were offered for what would be considered to be acceptable results. Local communities and nuclear opponents must be given assurance that even such preliminary site work will be held to the highest possible levels of technical achievement and intellectual honesty. This process must, of course, work both ways: governments cannot be held to specifically defined criteria before the first few bore holes are even completed.

The third regulatory problem of peer review and external credibility has already been mentioned. Public acceptance of both national and site-specific RWM programs must be based on the credibility of technical results. Evidence must be openly solicited and carefully considered. Hearings such as those at Windscale in the United Kingdom and on those radioactive waste management in the 1970s by the Nuclear Regulatory Commission in the United States may have been at least somewhat counterproductive, given the perceptions of many of a divergence between the apparent weight of evidence and the outcomes.

Fourth is the complex problem of coordinating various developmental and regulatory programs. Despite the need for the autonomy of regulatory bodies, coordination and priorities must be imposed by the government. Without such leadership, several regulatory agencies in a single system do not appear to be capable of producing credible, timely, and consistent results.

Fifth is the politicization of regulation. Difficult decisions that should be made by the president or parliament cannot be passed off on the regulators. This further aggravates the current state of regulatory overload. Although this general problem is not confined to the United States, the NRC appears to have a particularly unreasonable set of specific responsibilities in the highly controversial and political area of nuclear exports, nonproliferation, and foreign policy.[12] This also relates directly to the problem of coordination, since difficult decisions like those of nonproliferation policy cannot effectively be made in several different agencies.

The final problem is the demand on regulatory bodies to help resolve intergovernmental problems, especially between the federal, state, and local levels. Whereas this function is not generally important in France, the Soviet Union, the United Kingdom, and the less-developed countries, it is increasingly influential elsewhere. When the required general guidance is provided by executive and legislative arms of governments, this appears to be an appropriate use of regulatory systems. If regulatory agencies act with reasonable speed and consistency, they seem to help assure local and state governments that adequate legal and technical standards are being met.

COMPLEX AND COMPETING RELATIONS AMONG LOCAL, STATE, AND FEDERAL LEVELS OF GOVERNMENT

The consultation process between federal governments and states over RWM issues in several countries demands a very delicate balancing of federal and state powers. Austria, Canada, Japan, Switzerland, Sweden, West Germany, the United Kingdom, and the United States have already encountered problems over transportation and test drilling operations. Even in France, operations such as transporting Japanese spent fuel to La Hague have encountered widely publicized local opposition.

Discussion of the RWM issue is greatly simplified by use of the framework discussed above (table 2.1) which outlined principal issues, actors, and time periods. Regardless of how overall responsibilities are ultimately assigned for these three issues, the states are important actors in every case.

The state or provincial governments of several countries are involved in many elements of executive branch RWM and spent-fuel programs. In some cases, such as tightening standards for low-level waste disposal facilities, deciding on sites for expanded or new spent-fuel storage plants, and siting test facilities for high-level waste disposal, the state or provincial governments have an immediate and central role.

National and state level political party organizations are relatively neglected aspects of intergovernmental relations in RWM issues. Parties serve at least two functions here: shaping government policies and providing for debate and exchange of views. Although debate within and among the parties will not be important in the Soviet Union and the less-developed countries, it has already exerted influence in France, Sweden, and West Germany, and it could become important during the 1980s in Canada, Japan, the United Kingdom, and the United States.

It may be unfortunate that to date the political parties seem to have only contributed to the intensity of an already polarized debate over nuclear energy, especially in Sweden and West Germany. This is partially because very high political stakes have been involved for this issue in both cases. But there are signs that the traditionally pronuclear stances of political parties in Europe, for example, are now being modified. This may provide opportunities for less intense debate and exchange in other countries, which could help release intensifying political pressures, settle disputes, and deal with federal-state relationships. One clear but very difficult element of success is separating, whenever possible, disputes over general nuclear issues from specific questions of RW and spent-fuel management.

If one correlates the health of the nuclear industry with a crude plot of public access to RWM decision making versus centralization of governmental structure (see fig. 2.2), at least one trend emerges. It is those countries with relatively decentralized decision making and systems which are particularly open (at least for information) in RWM that also currently face the worst nuclear industrial prospects. This includes Sweden, West Germany, the United States and Canada. Despite this situation, it may be that these countries have reached, or will soon reach, closure on at least some of the important nuclear policy issues. Any reasonable level of compromise may partially defuse the issue and create a climate for limited, but predictable, levels of industrial activity.

PUBLIC POLICY RESPONSES

Linkages between Nuclear Power and Radioactive Waste Management

Many people from industry see the Swedish example as showing the severe difficulties that arise if industry and government programs in nuclear power production are not segregated from radioactive waste management. Environmental groups in most countries have in fact very effectively achieved their goal of holding nuclear energy hostage to the radioactive waste problem, thereby

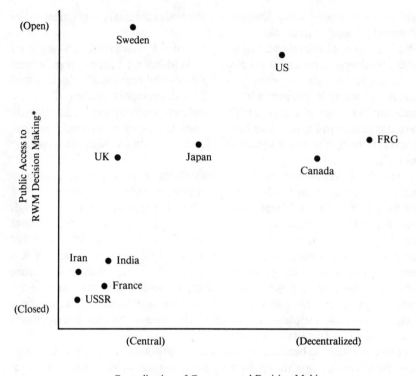

Figure 2.2   Public access to RWM decision making and centralization of governmental structure for ten countries

forcing governments to proffer planned solutions or even to attempt some type of expedited demonstration. There are clearly broad social concerns with the unsolved problem of radioactive waste in light of an ever-growing nuclear industry, and there are clear goals on the part of many environmental organizations to maintain linkage between the two issues. There are also long-standing fears, which have now been confirmed on the part of the industry, utilities, and various national governments, that lack of visible progress in solving the waste problem would constrict or even terminate the growth of the nuclear industry.

Linking nuclear power production with RWM does have a positive side in that it makes explicit the incentives to solve the RWM problem. On the other hand, proceeding too quickly toward disposal options could result in failure of both the technological and sociopolitical aspects of the problem. What is needed is a carefully constructed linkage with very discreet RWM steps in processing and disposal and the means of testing at each stage.

A socially and technically acceptable set of RWM steps which can be tied to progress in nuclear power could be developed without holding new nuclear plants hostage to the final demonstration of disposal technologies. Clear thinking demands separating radioactive waste management in the interim stage from final disposal. Linkage should be sufficiently strong for developing a credible radioactive waste management plan, but sufficiently weak between licensing new power plants and demonstrating the final steps in the disposal process so as not to destroy prematurely the nuclear industry. What is needed is a consistent, coherent, and historically verifiable set of policies which will lead to solving the final waste and spent-fuel management problems. Each country will have a somewhat different approach based on its unique cultural and political environment.

A Two-Stage Process for Radioactive Waste Management

A two-stage process could provide a framework for developing national programs in radioactive waste management, given the very wide disparity that exists between the availability of final disposal options and the immediate needs for interim measures such as spent-fuel management and the establishment of national radioactive waste management strategies. The first stage in this process is implementation of the interim measures over the next one to two decades. The second stage is the development of permanent geological disposal, based on a program of discreet schedules and visible progress toward this final goal.

The first step requires reaching agreement, at least at national and state levels, that the demonstration of final disposal is not linked to the continued use and expansion of nuclear energy. West Germany and Japan provide strong negative examples of why a direct linkage between final full-scale disposal or final full-scale disposition arrangements should not be linked to immediate reactor licensing requirements. The results are not constructive for the broader social objectives of safe disposal of radioactive waste.

The second component does require successful completion of the steps established in the permanent disposal plan, including meeting milestones on time, before increasing (or perhaps, in some cases, continuing) reliance on nuclear power. The partial decoupling of these two stages is designed to minimize risk to public health and safety without explicitly promoting or halting nuclear energy. Implicit in this approach is that governments must maintain some degree of balance in energy research and development programs as an insurance policy in case of a completely stagnated nuclear industry.

The role of national regulatory agencies, such as the Nuclear Regulatory Commission in the United States, is an essential part of this two-stage process.

Special attention must be given to the unrealistically high expectations placed on regulatory agencies, given their available public support, funding, technological expertise, and other limitations. National regulatory agencies must generally set the requirements for progress toward the final geologic disposal goal which, if met, will allow continuation of the interim steps and the nuclear industry. They must specifically establish the number of geologic disposal sites and media that should be submitted for licensing, which in turn influences the design of the research, development, and demonstration program that leads to this licensing process. The regulatory agencies also function as a link in the process of bringing together different elements of the federal government, including groups that are active in nuclear policy issues.

The national regulatory agencies must also ensure that a broad systems approach is applied to the process of establishing the waste disposal facilities, including all the interim management steps and their social and economic implications. This involves requiring a methodology for careful comparative assessment of the risks of storage as opposed to disposal, and of land as opposed to sea disposal, in current and future programs. The regulatory agencies also have a central responsibility to help link the federal government with events and responsibilities in local communities and the states. Finally, the regulatory agencies should be a focal point in the process of framing a balance in emphasis among physical science, social science, and humanistic considerations.

NOTES

1. In the United States many communities and states have already banned or restricted the transport, storage, or disposal of radioactive waste. It remains a highly contentious and unsolved issue in the U.S. Congress. In 1979, for example, Rep. Richard Roach, (D-Mass.) introduced two bills to limit the on-site storage of spent fuel by banning both new storage pools and the expansion of capacity for old ones.

2. Carroll L. Wilson, "Nuclear Energy: What Went Wrong?" *Bulletin of the Atomic Scientists,* June 1979, pp. 15, 17.

3. Relatively large-scale and politically important public participation programs have also been conducted in the Netherlands and Austria; see Dorothy Nelkin and Michael Pollak, "The Politics of Participation and the Nuclear Debate: A Comparative Study in Sweden, the Netherlands, and Austria," *Public Policy,* Summer 1977.

4. There were also controversial questions of the balance of executive-congressional budgetary control and the degree to which interim spent fuel storage would delay, or even replace, reprocessing.

5. A small pilot reprocessing plant has been operated at Karlsruhe since 1971; see E. William Colglazier, "Potential Impact on the United States of Foreign Radioactive Waste Management Technical Policies and Programs," Center for Science and International Affairs, Harvard University, June 26, 1980.

6. For an example of the confusing disagreement in the U.S. scientific community over solidification technologies, see Luther Carter, "Academy Squabbles over Radwaste Report," *Science,* vol. 205, July 20, 1979, pp. 278-289.

7. The social, cultural, and political underpinnings of decision making in nuclear power and the fuel cycle are analyzed in Dorothy Zinberg, "The Public Response to Nuclear Energy," in Frederick C. Williams and David A. Deese, eds., *Nuclear Nonproliferation: The Spent Fuel Problem,* (New York: Pergamon Press, 1979).

8. See M. Lonnroth and W. Walker, *The Viability of the Civil Nuclear Industry,* International Consultant Group on Nuclear Energy (New York: Rockefeller Foundation and Royal Institute of International Affairs, 1979).

9. The PTB is a national laboratory in Braunschweig under the jurisdiction of the Federal Ministry for Economic Affairs.

10. This section is adapted from Jean-Claude Derian and Irwin C. Bupp, *Running Water: Nuclear Power on the Move in France,* The Keystone Conference on International Implications of Radioactive Waste Management, Keystone, Colorado, Sept. 15-17, 1979.

11. The NRC Procedural Rule (10CFR 60) adopted in early 1981 requires characterization at depth in at least three sites in two geologic media before DOE submits a license application for constructing the first high-level waste repository.

12. Further examples of this problem include the U.S. NRC GESMO hearings, the "confidence" hearings by the NRC on links of RWM to reactor licensing, and the Windscale hearings in the United Kingdom.

3

What Happened to the IRG? Congressional and Executive Branch Factions in Nuclear Waste Management Policy

Thomas H. Moss

If a truism were defined by the frequency of repetition, the statement that nuclear waste disposal presents an institutional, not technical problem, would surely qualify. Yet with nearly equal frequency, sometimes from the same individuals, one can hear the observation that there are a number of serious unresolved technical issues in waste management. The contradiction represented by the unrefined juxtaposition of these two ideas is at the root of many of the seemingly incredible number of false starts, dead ends, political disasters, and technical mix-ups which have marked our nation's nuclear waste programs. It is one of the reasons that, despite massive investments of more than a billion dollars in the last three years, a sense of drift or paralysis has still characterized our nuclear waste disposal efforts. More specifically, it is an important part of the explanation for the failure of the Congress to reach any coherence in its own approach to the problem, and for the debilitating loss of momentum and credibility of the Carter administration's initially well-grounded Interagency Review Group (IRG) nuclear waste disposal initiative.

The confusion generated centers, in my view, around a failure to understand how strongly the technical and institutional questions interact. It is no surprise that there is a great deal of emphasis on institutional issues in seeking solutions to nuclear waste disposal problems. The seriousness and complexity of the state-federal and military-civilian relationships, among others raised by nuclear waste authority and responsibility discussions, cannot be minimized. Nonetheless, it is also critical to keep in mind why these institutional issues are particularly complex in this case. Largely, it is simply because the technology is so unfamiliar, and seems to raise a whole new order of risk-assessment and technical-performance problems. For the institutionalist to forget these under-lining technical matters is to exacerbate all of his other problems. In fact, concentrating a high fraction of resources on technical progress, and building incrementally a body of practical technical experience and familiarity, may well be the best way to ease institutional tensions and reach needed institutional compromises.

On the other hand, even with the technocrat's optimism in abilities to find complete scientific and engineering solutions, it is critical to realize that public confidence must be built up stepwise, through careful demonstration of the performance of system components and a willingness to submit those demon-strations to full institutional and public scrutiny. Pressing ahead too rapidly with technical development, without paying sufficient attention to the con-fidence-building functions ideally played by institutional review and to the need for new institutional compromises where new technological ground is broken, can lead to fatal weakness at future times.

Both administration and congressional efforts in nuclear waste questions have been weakened by a tendency to ignore one or the other of the compo-nents of the technical-institutional equation. The inability to integrate both aspects of the problem, and the tendency to concentrate debate and resources on one to the neglect of the other, have led to the failure of many otherwise promising approaches.

THE TECHNICAL ISSUE

The technical challenge by itself appears less than overwhelming simply be-cause it does not represent the kind of "unknown" that the scientific commu-nity associates with uncrossed frontiers. The long-term thermal and radiological characteristics of the waste are well understood, and a variety of barriers ranging from initial encapsulation through deep geologic storage seem redundantly to promise long-term isolation from the biosphere. Many of these have been designed, some built in prototype, and a few tested on limited scale at various locations around the world.

Yet clear technical uncertainties remain. The detailed physical and chemical behavior of many of the geological media under long-term heat and radiation stress cannot be fully predicted. Similar questions exist concerning the encap-sulation or immediate containment cylinders used. The geochemical and bio-chemical transport mechansism of all the various releasable isotopes cannot be completely documented under the widely varying conditions encounterable.

The methodology of risk assessment is at a stage where the combination of uncertainties can only be crudely compared with other societal hazards. Most of these problems do not seem fundamental. However, it is also true that the systematic analysis, and more importantly, empirical test data for the multitude of interacting effects, have not been developed and synthesized.

Experience also warns of extensive systems-engineering problems likely to be encountered before a smoothly running operation can be achieved. Transportation systems are certain to suffer at least occasional delay and confusion. The first lost or disabled vehicles, in unfamiliar locations, with operators or local officials forgetting or misapplying untried procedures, will inevitably cause public anxiety and controversy. Handling, emplacement, and ventilating systems will certainly at some time be erroneously operated or suffer breakdowns. The very redundancies built in for safety reasons may prove, in operation, to generate risks, as they have in certain other complex systems. That is, the apparent security of redundancy and lack of precise definition of critical factors may lead to operational neglect of some elements of the system under a false assumption of protection by others.

More fundamental will be the issue of designing the system for assurance of containment over an acceptably long period of time. The argument over whether that time is hundreds, thousands, or tens of thousands of years is not strictly at issue. In this society, where seven-to-ten-year depreciation times and thirty-year building life cycles are typical, even records of two generations have to be researched as if they were ancient history. With this kind of common practice, even the minimum estimate of needed radioactive waste control time will break new ground in systems design. Further, no matter what general principles of geologic containment are developed, site-specific assurances will have to be sought in detail to show skeptics that the particular site chosen will not prove to be the inevitable exception to whatever general rules seem to exist.

The fact that this set of problems is vexing but still seemingly solvable is the reason for the puzzling juxtaposition of declarations like that from the IRG, "Successful isolation of radioactive wastes from the biosphere appears technically feasible for periods of thousands of years", with equally authoritative statements that there are serious technical problems in nuclear waste disposal.

LINKAGE BETWEEN THE TECHNICAL AND INSTITUTIONAL ISSUES

The "minor" uncertainties in the "almost solved" technical problem are the major cause of the apparently "almost unsolvable" institutional issues. Licensing procedures cannot be designed around a mere set of intentions to try a number of detailed processes. Courts are unlikely to rule that the public safety is adequately protected on a basis of little more than strong technical intuition that one or a combination of these will be workable. State officials and politicians cannot be expected to commit their prestige and leadership to acceptance of a system which can only be described to them in the loosest of terms. The public itself cannot be expected to simply accept assurances of smooth future functioning of a planned system when the only palpable examples have been leaky tanks, bankrupt "walk-away" private operators, and protracted legal disputes in which federal, state, and private officials maneuver to *avoid* taking responsibility for protecting public health and safety. Neither can the public or the political-economic system be expected to define an equitable system of allocating and compensating for the negatives for nuclear waste responsibility where there is no experience with the magnitude or character of those negatives. Each of these issues has been acutely exacerbated by the history of unfulfilled promises in nuclear issues, and, more profoundly, by the obviously deep-seated public sensitivity to risk questions associated with nuclear power and ionizing radiation.

Unfortunately, many of those who recognized the seriousness of the institutional issues have failed to put sufficient stress on their underlying technical cause. Because of the obvious regulatory chaos and lack of consensus on institutional mechanisms to allocate, regulate, and manage nuclear waste responsibilities, a great deal of debate and effort has been centered on defining the institutional framework. This has often led to profound political-philosophical debates over what amounts to hypothetical and academic issues, which greater experience and familiarity might have reduced to matters much more obviously related to common or historical experience.

On the other hand, many of those who have felt sanguine about the eventual prospects of effective technical solutions to all remaining problems have ignored the fact that the regulatory and political process is best based on experience, not theory. In attempting to ignore the technical issues of the waste problems by simply deferring investment in proving their solution to a later time, or by expecting the institutional mechanisms to move rapidly to an operational stage in the absence of documented experience with proposed techniques, they have tended only to set the stage for further setbacks.

CONGRESSIONAL APPROACHES

Congressional initiatives in nuclear waste have fully displayed these myopia, often exaggerated by the fragmented committee jurisdictions. One class of legislative thinking has concentrated on putting Department of Energy (DOE)-administered pilot or demonstration programs on rapid design and construction timetables. Until relatively recently, these were often associated with strong assumptions that it was a useless or diversionary waste of time and capital to question the oft-described strategy of glassifying liquid reprocessing waste, followed by disposal in steel capsules in bedded salt. These "go-ahead" programs have been designed for and by those who are satisfied that solution to all technical problems is at hand and have often been accompanied by declaration that "it is time to get on with the job." They have tended to minimize attention to alternatives and to ignore or seek override of the processes which have evolved in recent years for providing defined review procedures of new or dangerous technology development.

On the other hand, an entire conflicting genre of legislation has developed which focuses almost exclusively on findings, procedures, reviews, and hearings, often with associated deadlines to restrict or halt aspects of nuclear operations or licensing in the event of unsatisfactory outcomes. The extremes of this genre pay little attention to the programs, tests, and investments needed to provide the data to make the procedures meaningful. Most of this legislation has concentrated on roles for the Nuclear Regulatory Commission and the Environmental Protection Agency, often with undisguised distrust of DOE.

In addition to the philosophical poles represented by these ideas, in other cases legislation has been characterized by approaches determined primarily by only thinly veiled jurisdictional fixations. Committees with NRC jurisdiction have written legislation entirely subsumed in licensing, as opposed to technological, considerations. Research-oriented committees have focused on development programs that at best ignore, and at worst seek to evade, procedural safeguards. Committees dealing with the intra- and intergovernmental jurisdiction have concentrated single-mindedly on defining interagency councils and state-federal relationships. With nuclear waste responsibilities dispersed among several committees, and much more concern among them for expanding and protecting jurisdiction than for establishment of communication bridges and cooperation, there has been little progress in congressional approaches toward offering comprehensive and broad-base solutions.

In both cases, hidden or not-so-hidden agendas related to promotional or negative attitudes toward nuclear power have often obviously lain behind both the prose and substance of the nuclear waste policies proposed. Despite a generally professed belief that an effective nuclear waste policy was important to both sides of the broader nuclear debate, neither side has been able to resist the temptation to use the waste policy discussion as an additional arena to maneuver for rhetorical or tactical advantage in the larger pro- and antinuclear debate.

Many of the narrow congressional initiatives have shown a certain initial flare of attention and political strength, but very little staying power. Starting in friendly subcommittees, they are heralded and enthusiastically reported upward. In full committees questions may begin to appear, but it is often only when reported for full House or Senate action that the Congress has been faced repeatedly with seemingly irreconcilable differences of approach. It has also been at this point that the administration has normally made its own assessment of the seriousness of congressional intent. An indicator of that seriousness would certainly be a willingness among congressional factions to set aside disputes and play a coherent role in formulating national policy. Once again, until very recently that assessment would have had to be that Congress was not able or eager to play a serious partnership role with the executive branch in seeking solutions to a national dilemma. The absence of *coherent* congressional pressure on the executive often has had the effect of no congressional pressure at all, and has left nuclear waste disposal at a policy-priority level within the administration far lower than the impact of the issue on Congress would suggest.

CARTER ADMINISTRATION EFFORTS—THE "IRG"

In some ways the Carter administration seemed on the brink of making important progress toward integration of the various points of view via the "Interagency Review Group" (IRG) process it set in motion in 1977. Within the administration were key policymaking figures encompassing a very wide range of opinion on the subject. It was thereby possible for the executive branch to sense vividly within itself the tensions that needed to be resolved, as well as to recognize the compromises necessary to resolve them. Effective leadership was allocated to the effort so that the group was convened under an authority sufficient to overcome the centrifugal tendencies inevitably found in a group of federal agency representatives. Progress was made on several points long-stressed by outside critics of the nuclear waste program. Most important was the consensus forged that it was desirable to explore possibilities in several geologic media—instead of being technologically and politically trapped into defending a fixed position that bedded salt was the optimum geologic medium. The need was recognized for comparative chemical and physical experimentation in these media, with the aim of duplicating conditions to be faced in a repository. An orderly and logically incremental experimental program was defined. The consensus position of systematic and sustained development of data for several strategy options marked such a great release from the dogmatic defensive positions of the previous few years that it constituted a major step forward.

Some delicate political/philosophical compromises were formed as well. All participants were willing to endorse a notion of consultation and concurrence for states—in which the principle was accepted that state officials were to be closely involved with all stages of federal planning, without attempting to define the exact timing or locus of a yes or no decision on proceeding. Similarly, civilian licensing for some additional military waste facilities was recommended, but final decisions were deferred pending further study.

When the draft report of the IRG to the president was first published in the

the late fall of 1978, it was clear that not all participants or outside observers would be entirely satisfied. There was nervousness that new concepts had been opened up but not sufficiently limited. In other quarters there was dissatisfaction that the concepts opened had not been codified and given an absolute precedence over the old. Still, however, it was widely recognized that the IRG process had brought forth a policy framework with the possibility of a much broader political and technical support base than had been available to the nuclear waste program for many years. The president was in the enviable position of having his own bureaucracy relatively well disciplined to the consensus, of having disassociated himself from many of the political failures of the past, and of having the freedom to move to a policy of technical flexibility and political pragmatism. He could anticipate the accrual of all the political benefits of leadership in an area where drift had been chronic.

Perhaps more important, from the president's point of view, was the prospect of controlling policy initiative in this important and politically sensitive area. Given the incoherence and inconsistency of the set of various congressional approaches, the IRG framework had the potential of enabling the administration's proposals to become the benchmark for debate and the definitive statement of issues. The IRG proposals were the only complete and reasonably balanced set on the public agenda. In addition, even the crudest political interest of the various congressional proponents of much narrower strategies could have been well served by the IRG synthesis. Its existence alone provided a focal point for compromise among the strongly held congressional points of view, without the need for initial sacrifice by any individual.

With this propitious beginning in November of 1978, the inevitable question raised in looking back is, "What happened to the IRG?" My own interpretation is that the delay of more than a year in administration endorsement of its own carefully crafted initiative, the loss of policy momentum and congressional and public credibility, and the fragmentation of the compromise were the result of a basic error of the president's most immediate advisory group on this issue, largely within the Executive Office and DOE. Specifically, the failure was due to a dogged determination of factions in this group to raise and re-raise their points of view in terms far more specific than was needed. This activity was unfortunately most intensive *after* the results of the long interagency deliberations had been synthesized and submitted to the Executive Office. There was a continuous effort to force the president into a position of referee among staff philosophical positions. His time, attention, and political capital were demanded for questions which need not have been resolved in order to move forward with an effective national policy. Given the inevitable pressure on those scarce presidential resources, the inevitable result of demanding access to them was delay. The delay fed back on the system and provided more opportunity for heightening tensions, and greater temptation to press for individual philosophical agenda in a debate which, in principle, had already forged itself into a viable compromise.

The specific issues over which the IRG process later foundered were that of state veto of waste respository siting and that of extending licensing of military facilities.[1] Both of these issues underlay factional arguments over the continuation of planning for the military transuranic (TRU) waste isolation facility (WIPP) in New Mexico. Had the general IRG statements of broad compromise and continued concern for these issues been promptly endorsed by the president, the many consensus initiatives in more detailed areas could have been immediately implemented under the prestigious banner of a new and coherent federal policy. However, continuously seeking more definite presidential commitment to the poles of the debate on these two fundamental questions appeared to preoccupy the Executive Office even at the final decision points, and diverted attention from the need to consolidate the gains of consensus on less fundamental issues. Congress was diverted too, I believe, in its reception of the IRG product. Instead of taking the opportunity of using the consensus positions of the IRG as a way of reaching an acceptable compromise on an issue that had thus far been a chronic irritant and embarrasment, its own reactions were largely couched in terms of pressing for more definite commitment, one way or another, on the two key policy questions raised above. This was, of course, despite its own failure to obtain policy consensus on these.

THE BREADTH OF THE INSTITUTIONAL QUESTIONS

The arguments above are not meant to minimize the importance of the state-federal and military-civilian relationships in nuclear waste policy. In fact, the point is made more to emphasize and stress them—with the implication that they are broader and more consequential than the nuclear waste questions which currently bring them to the surface. In times when resources have been scarce and economic interests divergent, or when there have been "negatives" to allocate among states or regions, the state-federal relationship has always been a source of tension in this country. Emerging issues of water and energy resource scarcity, chemical waste disposal, land use and preservation, acid rain and other "air-shed" allocation questions, and many others will all provide further and possibly more demanding tests of this relationship.

In this broad context, it was a mistake to link presidential action and endorsement of the technical strategy of the IRG nuclear waste disposal program to a rigorous new definition of the state-federal role and a final "de jure" designation as to which party has a veto or preemptive right.[2] The technical development strategy was practical and implementable immediately. However, the final site selection for a repository was at least two presidential terms away. There was no need for this president to use his political resources to take a definitive stand on an issue that had been viewed in terms of loose compromise for generations. Certainly there is some merit in looking ahead institutionally as there is in planning ahead technically. However, there is also room for institutional experimentation and evolution, especially when the dimensions of the institutional problem are not yet clear. The president should not have been enmeshed in a policy debate aimed at cutting short that experimentation and evolution by forcing a premature decision on federal recognition of a state veto in the starkest terms. Even had that been his inclination, it would be presumptuous to believe that the sitting president, the sitting Congress, and the sitting state officials involved at the actual time of site selection would not be obligated to reconsider and make the final determination on the role of the parties.

Thus pressure on the president from his own staff, from outside groups, and from the Congress to make more precise the IRG statement of "consultation and concurrence" was truly misdirected. The structural nature of the repository, the mechanics of its operation, and even the geological media in which it is to be located were not yet defined. The attitudes of state politicians were indicative of the fact that resolution of state-federal question was being prematurely forced. Governors gave mixed signals as to whether they even wanted to be put into the position of having a state veto, and for good reason. Being trapped into a rigid position on a hypothetical, ill-defined, or prematurely raised question is a classic political mistake which wise state political figures were anxious to avoid. That the president's staff persisted in forcing such questions into the final stages of the process of preparing the presidential statement on the IRG recommendations indicated a foolish willingness to squander political capital and leadership potential on the resolution of intra-administration or intra-staff disputes.

The question of extending civilian Nuclear Regulatory Commission licensing to all military nuclear waste operations was not quite on the same broad historical and social scale, but nearly so. The nature of civilian review of military operations in general has been another persistent source of tension in our political system. In particular, civilian control of military *nuclear* operations has been especially sensitive ever since the wartime secrecy of the Manhattan project. The tension was reflected in the original drafting of the Atomic Energy Act of 1954, which specifically set distinctly different procedural ground rules for civilian and military nuclear programs. It has continued in all subsequent alterations of that act and was an important issue in long debate in the legislation changing the Atomic Energy Commission into a broad-based Energy Research and Development Agency (ERDA).[3]

Unfortunately, in this area too, the administration acted as if it were unconscious of some of the broader issues and long-standing sensitivities of the military-civilian nuclear relationship. Just one example was the failure to properly draw the House and Senate armed services committes into the decision-making process. Even though these groups shared with the administration a strong interest in rapid technical and institutional resolution of nuclear waste

questions, relationships soon became adversarial. In early prepublication discussions of IRG progress, the administration surfaced the idea of a civilian
waste pilot facility added on in an undefined way to the planned military
transuranic (TRU) waste isolation facility (WIPP) in New Mexico. This raised
the specter, in armed service committee minds, of civilian licensing being
extended to a new class of military waste activity. That this specter was raised in
a nearly casual manner, without proper prior discussion of intent and intended
limits, led to extreme suspicion of all subsequent IRG ideas by the political
constituency of the military program.

This suspicion was further fueled by subsequent action after the IRG was
sent to the president for final decision. Once again, even though the IRG group
had endeavored to strike a compromise position which the president could
endorse, factions in the administration and the president's immediate staff
continued to wage rear-guard battles to swing the presidential decision to their
side on the issue of whether or not to proceed with an unlicensed military TRU
facility (WIPP). In this case, too, an unacceptable and unnecessary choice was
being thrust on the president: to use his political prestige to try to resolve, in
definitive fashion, a tension that had been present in specific nuclear terms for
thirty-five years and in general terms for two centuries. There was nothing in
the IRG intent of realigning the technical goals of the civilian nuclear waste
program which required final resolution of this issue or which required tying
presidential action on the IRG report to a definitive position on the broader
matter. By persisting in involving the president and the IRG recommendations
in this broader dispute, presidential staff helped to insure the kind of delay of
presidential IRG endorsement which was so damaging to its credibility.

WHAT SHOULD THE PRESIDENT HAVE DONE?

I think true leadership would have been read into a prompt IRG endorsement
statement which straightforwardly indicated the president's perception that the
state veto issue was in fact one of a class of modern technology-related tests of
the notion of federalism. He could have pointed to a national choice between
accepting the consequences of complex organization and technology, or of
preserving the purest historic notions of state individuality. Most importantly,
he could have pointed out the obvious: that whatever his or contemporary
political preference, the question was theoretical with respect to repository
siting at this time. Whatever political statements were issued now, or whatever
technical explorations were undertaken, it would be a future president, a future
Congress, and a future set of state politicans in office at the time of a siting
decision who both should and would in fact have the final decision. Similarly,
whatever his views on licensing military waste operations, he could have
proposed that the WIPP transuranic (TRU) defense waste facility go ahead on

a small experimental scale as a part of a coherent research and development program and as a valuable opportunity to gain operational experience. The licensing of full-scale military TRU waste operations could similarly be taken up the in due course as the separate political question it is, in the long-standing context of continuing evolution of regulatory thinking with respect to military nuclear activities, and not as bottleneck on formation of a respectable waste program.

With both of these institutional questions defused by placing them in their properly broader context, full commitment and political support could have been directed to a program of building the waste-handling knowledge base, of providing flexible alternative strategies to deal with unforeseen technical developments, and of strengthening public and political confidence through cautious buildup of actual experience in small-scale test and pilot operations. In fact, the president's final statement, more than a year after it was initially expected, was not far from the original IRG recommendations. The controversy allowed to develop over the WIPP facility licensing question forced abandonment of that initiative even on an experimental scale, but the most important of the technical strategy and consultation and concurrence concepts were preserved. The State Planning Council established by the president as an elaboration of an IRG discussion could turn out to be an important institutional innovation with utility far beyond questions of nuclear waste. The administration's effort and basic approach in this context commands admiration and respect. However, it also makes all the more tragic the fact that unnecessary and inappropriate factional controversies were allowed to dissipate momentum and leadership.

In summary, the president was presented with political and philosophical choices which did not have to be made for successful renewal of progress toward a nuclear waste management system. The resulting controversy and concomitant delay in presidential action on the IRG left a vacuum where leadership was needed and tended to dissipate the broad consensus on technical issues which had been so much to the credit of the president's initiative. It projected future action into the uncertainties of an election year, and probably undermined prospects in the Congress of building coalitions of joint action around the consensus areas of IRG findings. By pressing more eagerly to win policy debate rather than to form policy, factions in both the administration and Congress ensured further delay in reaching a nationally acceptable nuclear waste management strategy.

NOTES

1. The number of potential media or sites to be explored and developed was sometimes at issue, both in principle and in a budgetary sense.

2. The basic statute governing federal-state relations is the Atomic Energy Act of 1954. This gives the federal government nearly complete control of regulatory authority except in narrow areas where it is specifically delegated to the states. Such delegation has not occurred for questions involving high-level or long-lived transuranic waste. The original federal preemption is largely still in force, though some subsequent statutes have explicitly involved states in management of some classes of radioactive material and have required federal-state notification or consultation mechanisms. A more thorough discussion of this point is found in Chapter 4 of this volume.

3. The Energy Reorganization Act of 1974, which created the Energy Research and Development Administration (ERDA, later evolving to the Department of Energy) and the Nuclear Regulatory Commission (NRC), specifically gave to NRC licensing authority over *high-level* radioactive waste disposal whether originating from civilian or military programs. Operations or disposals involving transuranic (TRU) waste from defense operations are not, however, required to be licensed under present law. NEPA (National Environmental Protection Action) provisions, requiring environmental impact statements, do apply as they do to military operations generally. This level of civilian authority does not have, however, anything approaching the stringent implications of a licensing procedure.

4

Federal-State Conflict in Nuclear Waste Management: The Legal Bases

Harold P. Green and L. Marc Zell

The management, storage, and disposal of America's vast and ever-increasing inventories of commercial and military radioactive wastes is perhaps the most crucial and complex issue confronting both advocates and opponents of nuclear power today.[1] Ironically, despite the vital importance of this issue to the future of nuclear energy, the response of the federal government has been characterized by delay and indecision which some regard as ineptitude and political mismanagement.[2]

Troubled by the technological and institutional uncertainties apparent in federal nuclear waste management activities, and concerned over rising public opposition to the siting and construction of radioactive waste repositories within the past three years, a number of state and local governments have enacted legislation designed to bar or severely restrict the transportation, storage, or disposal of nuclear wastes within their territorial boundaries.[3] While there appears to be general agreement that state involvement in any prospective national waste management program will be critical to the success of such a program,[4] the legislative alternatives now being pursued by many states may well be unconstitutional insofar as they purport to regulate federal waste management activities or to the extent that they otherwise intrude impermissibly upon the existing federal statutory and administrative regime.

In this chapter we will examine the present legal bases, or lack thereof, for the exercise of state regulatory power over the storage, disposal, and transportation of radioactive wastes. We emphasize at the outset that the *legal* question of whether states and localities currently possess constitutional authority to enter the field of nuclear waste management is wholly distinct from the *political* issue of whether states *ought* to exercise authority in this area. It is solely with the former problem that this chapter is concerned. Ultimately, the responsibility for balancing the interests of the states and the federal government and allocating regulatory authority between them rests with the political branches of the federal government and with the Congress in particular.[5]

PRELIMINARIES

Statutory Overview

Before undertaking an analysis of the complex legal issues posed by the entry of state governments into the field of nuclear waste management, it would be instructive to examine briefly the allocation of regulatory responsibility under present federal law. The basic instrument of federal regulatory authority over nuclear matters and specifically over nuclear wastes is the Atomic Energy Act of 1954,[6] as amended by the Energy Reorganization Act of 1974:[7] the Department of Energy Organization Act of 1977;[8] and the Uranium Mill Tailings Radiation Control Act of 1978.[9]

Shortly after the end of World War II, Congress transferred exclusive authority for the control and development of nuclear energy to the now defunct Atomic Energy Commission (AEC), pursuant to the Atomic Energy Act of 1946.[10] Under this statute, the federal government enjoyed an absolute monopoly over all fissionable materials and related facilities.[11]

With the enactment of the Atomic Energy Act of 1954, participation by the private sector in the development of nuclear energy became possible for the first time. Under the 1954 Act, the AEC was vested with what appears to have been the exclusive power to license the possession, transfer, and use of "source," "special nuclear," and "by-product" material.[12] Although the 1954 act did not specifically mention nuclear waste, such materials clearly fall into each of the three categories enumerated in the 1954 act and are thus fully subject to the regulatory authority of the commission.[13] Before a license can be issued to a private party for any of these covered classes of nuclear materials, the commission is required to determine that the health and safety of the public would not be endangered.[14]

To clarify the respective roles of the states and the federal government with respect to the control of source, by-product, and special nuclear material, the Congress amended the act in 1959, adding a new section 274.[15] The 1959 amendment authorized the AEC to relinquish, by way of a so-called turn-over agreement, certain defined areas of regulatory jurisdiction over those three classes of nuclear materials which had theretofore been controlled solely by the commission. In particular, section 274(b) provides:

> [T]he Commission is authorized to enter into agreements with the Governor of any State providing for discontinuance of the regulatory authority of the Commission . . . with respect to any one or more of the following materials within the State -
> (1) byproduct materials as defined in section 2014(e)(1) of this title;
> (2) byproduct materials as defined in section 2014(e)(2) of this title [uranium mill tailings];
> (3) source materials;
> (4) special nuclear materials in quantities not sufficient to form a critical mass.
> During the duration of such an agreement it is recognized that the State shall have authority to regulate the materials covered by the agreement for the protection of the public health and safety from radiation hazards.

Under section 274(c), the commission is prohibited from delegating regulatory authority to states in certain areas, including in particular:

> the disposal of such other byproduct, source, or special nuclear materials as the Commission determines by regulation or order should, because of the hazards or potential hazards thereof, not be so disposed of without a license from the Commission.[16]

To date, twenty-six states have entered into turnover agreements with the commission.[17]

Section 274(k) explicitly preserves state regulation authority in nonradiation areas by providing:

> Nothing in this section shall be construed to affect the authority of any State or local agency to regulate activities for purposes other than protection of radiation hazards.[18]

The Energy Reorganization Act of 1974 abolished the AEC and in its stead created the Nuclear Regulatory Commission (NRC) to exercise all of the regulatory and licensing powers previously held by the AEC under the 1954 act. All the functions under the 1954 act were assigned to the Energy Research and Development Administration (ERDA). With respect to the waste management activities of ERDA, the 1974 act specifically vests NRC with licensing and regulatory authority over

> (3) Facilities used primarily for the receipt and storage of high-level radioactive wastes resulting from activities licensed under . . . [the 1974] Act;
> (4) Retrievable Surface Storage Facilities and other facilities authorized for the express purpose of subsequent long-term waste generated by [ERDA], which are not used for, or are part of, research and development activities.[19]

The 1974 act also established within NRC an Office of Nuclear Reactor Regulation with the function, among others, of evaluation of methods of "transporting and storing high-level radioactive wastes [from NRC-licensed facilities] to prevent radiation hazards to employees and the general public."[20]

Also created by the 1974 act is the NRC's Office of Nuclear Material Safety and Safeguards which is to assist ERDA in the development of safeguards for the protection of high-level radioactive wastes from "threats, thefts and sabotage."[21]

In 1977, Congress passed legislation creating a Department of Energy to which all of the functions of the short-lived ERDA were transferred.[22] Section 203 of the 1977 DOE act describes various duties to be exercised by the new department's assistant secretaries. Of special interest is section 203(8) which sets forth in substantial detail DOE's responsibilities in the nuclear waste management area:

(A) the establishment of control over existing Government facilities for the treatment and storage of nuclear wastes, including all containers, casks, buildings, vehicles, equipment, and all other materials associated with such facilities;

(B) the establishment of control over all existing nuclear waste in the possession or control of the Government and all commercial nuclear waste presently stored on other than the site of a licensed nuclear power electric generating facility, except that nothing in this paragraph shall alter or effect title to such waste;

(C) the establishment of temporary and permanent facilities for storage, management, and ultimate disposal of nuclear wastes;

(D) the establishment of facilities for the treatment of nuclear wastes;

(E) the establishment of programs for the treatment, management, storage, and disposal of nuclear wastes;

(F) the establishment of fees or user charges for nuclear waste treatment or storage facilities, including fees to be charged Government agencies; . . .[23]

The act specifically bars DOE from exercising waste management regulatory authority presently assigned to the NRC.[24]

Late in 1978, Congress for the fist time expressly addressed the specific question of federal-state regulatory responsibility in the field of nuclear waste storage and disposal. Section 14(a) of the NRC Appropriations Legislation for fiscal year 1979 directs that:

Any person, agency, or other entity proposing to develop a storage or disposal facility, including a test disposal facility, for high-level radioactive wastes, non-high-level radioactive wastes including transuranium contaminated wastes, or irradiated nuclear reactor fuel, shall notify the Commission as early as possible after the commencement of planning for a particular proposed facility. The Commission shall in turn notify the Governor and the State legislature of the State of proposed sites whenever the Commission has knowledge of such proposal.[25]

Furthermore, the NRC was directed to prepare a report on "means for improving the opportunities for State participation in the process for siting, licensing, and developing nuclear waste storage or disposal facilities."[26]

The year 1978 also saw the enactment of the first comprehensive legislation aimed at eliminating radiation hazards posed by a type of nuclear waste material known as uranium mill tailings, found in tremendous quantities in the vicinity of uranium and thorium mining sites throughout the West.[27] The 1978 Uranium Mill Tailings Radiation Control Act, which creates an elaborate mechanism for the management of mill tailings sites by the federal government in cooperation with states, Indian tribes, and persons who own or control inactive milling sites provides a program remedying existing radiation hazards at such sites. The 1978 act delineates in great detail the respective roles of the states and federal government in maintaining and regulating disposal sites for uranium mill tailings. The NRC is vested with plenary authority, in connection with the Department of Energy and the Environmental Protection Agency, for the regulation, oversight, and management of mill-tailings-related activities. State licensing standards must be equivalent to the extent practicable, or more stringent than, those of the NRC.[28] The act also requires that all federal disposal sites for uranium mill tailings be transferred, upon termination of licenses, to the federal government or the state in which the site is located for permanent custody.[29]

Finally, no fewer than ten separate proposals were introduced in the first session of the Ninety-sixth Congress for the purpose of authorizing affected states to veto, approve, or otherwise participate meaningfully in federal decisions relating to the siting, establishment, and operation of nuclear waste facilities.[30]

Classification of Nuclear Wastes and the Current Regulatory Regime

A proper analysis of the legal impediments to state action in the realm of radioactive waste management must necessarily take into account the system of waste classification currently applied by the NRC and other federal agencies with jurisdiction in this area. Under the current regulations, the pervasiveness of federal controls which governs to a significant degree the power of the states to regulate varies in accordance with the severity of the potential hazards posed to the public by particular classes of wastes. Thus, high-level wastes tend to be subject to far more comprehensive federal regulation and control than low-level wastes or uranium mill tailings, for example.

Although more refined systems of waste classification are described in the literature,[31] it is sufficient for present purposes to distinguish broadly between high- and low-level radioactive wastes. At present, there is no formal definition of the term "high-level radioactive waste" (HLW) in any of the applicable statutes or the NRC regulations. Appendix F to 10 C.F.R. section 50, which sets forth policy as regards the siting of fuel reprocessing plants and related waste management facilities, does define "high level *liquid* radioactive wastes" to mean

aqueous wastes resulting from the operation of the first cycle solvent extraction system, or
equivalent, and the concentrated wastes from subsequent extraction cycles, or equivalent, in
a facility for reprocessing irradiated reactor fuels.[32]

Liquid HLW may be solidified through sophisticated processes such as calcina-
tion and virtification.[33]

Until April 1977, the foregoing definition would have been adequate to
encompass nearly all HLW generated in the United States, since such wastes
were produced exclusively in nuclear fuel reprocessing. However, in October
1977, the Carter administration announced an indefinite freeze on all commercial fuel reprocessing.[34] As a consequence, spent reactor fuel (largely in the
form of irradiated fuel assemblies) has been rendered useless; that is, it is
waste, for all intents and purposes. The problem is that there is neither
statutory nor administrative authority for treating spent fuel as waste,[35] although there is little doubt that if spent fuel is deemed to be waste, it would
easily qualify as HLW, since it contains all the toxic and long-lived radionuclides found in high-level liquid waste.[36] In this regard, it is interesting to note
that the Marine Protection, Research and Sanctuaries Act of 1972 specifically
defines the phrase "high-level radioactive waste" to include "irradiated fuel
from nuclear power reactors."[37]

Current NRC regulations require that liquid HLW, produced at privately
owned reprocessing plants, be reduced to solid form and transferred to a
federal repository within ten years after separation of the fission products from
the irradiated fuel.[38] Disposal of HLW is forbidden on any land except that
which is "owned and controlled by the Federal Government."[39]

The regulations are silent as to the procedures to be followed in selecting sites
for federal HLW repositories. Nor are there any specifications for the design,
construction, or maintenance of such facilities. It is generally agreed that the
highly toxic nature of HLW, as well as the extremely long half-lives of the
nuclides contained in such wastes, mandates that HLW be isolated from the
biosphere for long periods of time. Whether such long-term isolation (permanent disposal) is technologically and institutionally feasible is the subject of
considerable debate.[40]

Transuranic wastes (TRU), which result primarily from spent-fuel reprocessing and nuclear weapons production, are those which contain transuranic
elements, for example, plutonium, americium, and neptunium in concentrations of greater than ten nanocuries per gram of material.[41] Unlike HLW, TRU
waste does not pose handling difficulties attributable to heat generation and
temperature increase.[42]

In 1974 the AEC issued proposed regulations which would have barred the
disposal of TRU at commercial burial sites for low-level wastes.[43] Although the
rule was never formally adopted, there is a de facto prohibition on the disposal
of TRU waste at all but one of the commercial burial grounds.[44] Since 1970

TRU wastes have been stored only at federal sites.[45]

It is unclear whether TRU wastes ought to be treated as high- or low-level wastes. At present, the regulations do not address the issue, although it has been suggested that, due to the long half-lives of transuranic isotopes, TRU wastes be disposed of in a manner similar to that for HLW.[46]

Low-level wastes (LLW) consist of a wide variety of discarded equipment, clothing processing residues, and other material which contains or may have been contaminated with radioactive materials.[47] Such wastes require little or no shielding; thermal cooling is not necesary, due to the relatively low concentration of radioactive materials.[48] Typically, LLW is disposed of through shallow land burial.[49]

Because of the diminished health and safety hazards associated with LLW, the management of such wastes has not been subject to strict federal controls.[50] While it has expressly elected to retain exclusive regulatory power over the storage and disposal of HLW,[51] the NRC has delegated its authority to license and regulate commercial LLW burial grounds to five states under the Agreement States program of section 274(c).[52] Whether licensed directly by the NRC or by the states under section 274, low-level wastes may be disposed of only on land owned by the federal government or by a state government.[53] Moreover, all state LLW regulatory programs instituted pursuant to a turnover agreement must be consistent with any applicable federal standards.

LEGAL IMPEDIMENTS TO STATE REGULATION OF NUCLEAR WASTE MANAGEMENT

State law which flatly prohibits the establishment of nuclear waste repositories within the situs state may be subject to constitutional challenge on a number of independent theories. First, to the extent a state purports to regulate a federal activity carried out on federal property (e.g., HLW disposal), the federal activity would be beyond the power of the state to regulate under constitutional principles of intergovernmental immunity. Secondly, if a state statute conflicts with applicable federal law or otherwise operates to frustrate the policies and objectives of Congress, the state legislation may be invalid by virtue of the Supremacy Clause and the judicially refined doctrine of federal preemption. Finally, state statutory provisions which close the borders of the state to interstate commerce originating outside the jurisdiction are subject to challenge under the Commerce Clause. A discussion of each of these concepts follows.

Intergovernmental Immunity

Well over a century and a half ago, in the celebrated case of *McCulloch* v.

Maryland,[54] Chief Justice John Marshall laid the doctrinal foundations for the principle of intergovernmental immunity whereby the states are forbidden to interfere with or obstruct the activities of the federal government, be it through taxation, regulation, or otherwise. At issue in *McCulloch* was the validity of a tax levied by Maryland on a local branch of the Bank of the United States, an instrumentality of the federal government established by act of Congress. Striking down the state tax, the Supreme Court, per Justice Marshall:

> If we measure the power of taxation residing in a state, by the extent of sovereignty which the people of a single state possess, and can confer on its government, we have an intelligible standard, applicable to every case to which the power may be applied. We have a principle which leaves the power of taxing the people and property of a state unimpaired; which leaves to a state the command of all its resources, and which places beyond its reach, all those powers which are conferred by the people of the United States on the government of the Union, and all those means which are given for the purpose of carrying those powers into execution. We have a principle which is safe for the states, and safe for the Union. We are relieved, as we ought to be, from classing sovereignty; from interfering powers; from a repugnancy between a right in one government to pull down what there is an acknowledged right in another to build up; from the incompatibility of a right in one government to destroy what there is a right in another to preserve.[55]

One hundred years later in *Johnson* v. *Maryland,*[56] the Supreme Court invalidated the application of a Maryland driver's license statute to a federal postal employee driving a government-owned vehicle in connection with his official duties. Relying on *McCulloch,* Justice Oliver Wendell Homes, Jr., explained:

> It seems to us that the immunity of instruments in the United States from state control in the performance of their duties extends to a requirement that they desist from performance until they satisfy a state officer, upon examination, that they are competent for a necessary part of them, and pay a fee for permission to go on.[57]

Under the rationale of *McCulloch,* as applied in *Johnson,* a state is powerless to regulate a federal instrumentality in the absence of congressional authorization where the federal entity or activity has been established pursuant to or in furtherance of a power or powers specifically enumerated in the Constitution as belonging to the national government.[58] As stated earlier, the federal government's statutory authority to regulate the storage and disposal of radioactive wastes emanates from its power under the Atomic Energy Act of 1954 to license the use, transfer, and possession of by-product materials, which, in turn, is expressly predicated upon congressional power to regulate interstate and foreign commerce and to provide for the common defense.[59] Thus, inasmuch as federal authority over nuclear wastes is grounded in an enumerated power delegated to the national government under the Constitution, federal agencies and instrumentalities engaged in nuclear waste management operations enjoy an immunity from state regulation under the rationale of *McCulloch* and *Johnson*. Consequently, until such time as Congress acts to waive federal immunity, state laws which proscribe the siting of nuclear waste repositories anywhere within a state's territorial boundaries would be invalid as applied to federal property. This is particularly so with respect to HLW disposal sites which must be located extensively on federal land under the NRC regulations.[60]

It is conceivable, on the other hand, that HLW which is provisionally stored at privately owned facilities pending final transfer to a federal installation might not be protected from state control on the theory of intergovernmental immunity. With respect to such facilities, the state legislation would impact on the private entity and not the federal government, its property, or its instrumentalities, as was the case in *McCulloch* and *Johnson*.[61]

By the same token, any state action which purported to affect HLW before its delivery to a federally owned and operated waste repository might be deemed ineffective on the grounds that "a State is without power . . . to provide the conditions on which the Federal Government will effectuate its policies."[62] A state regulation which prevented private owners of commercial power facilities, for example, from shipping their high-level wastes to a federal repository would clearly frustrate the federal policy mandating the delivery of all such wastes to federal facilities. Under these circumstances, the federal immunity from state regulation may well extend to private owners and shippers of HLW.

Finally, as to LLW, since all but one commercial burial grounds are either licensed and regulated by the states pursuant to the Agreement States program, no issue of intergovernmental immunity would arise in this regard. However, those DOE LLW disposal facilities which are situated on federal lands would, of course, be insulated from state controls under the doctrine.[63]

Preemption

While the lack of state authority to regulate in any manner the storage of a disposal of HLW on federal reservations is plain under principles of intergovernmental immunity, that doctrine would not appear to effect state action with respect to the management of low-level wastes and HLW not located on federal property. It is here that principles of federal preemption come into play.

That Congress may properly legislate over the field of nuclear waste storage and disposal under the commerce and defense powers granted in Article I of the federal Constitution is not open to doubt.[64] By the same token, the states have traditionally wielded broad powers to enact legislation for the protection of the public health and safety powers reserved to them by the Tenth Amendment.[65] Absent the existing complex scheme of federal nuclear energy legislation and administrative regulations, there would be no question that the states could, under their police powers, safeguard the public against the grave hazards posed by radioactive wastes.

Where, however, the police powers of the states are exercised in such a way as to overlap or conflict with federal legislation enacted pursuant to a power vested in the national government, the Constitution's Supremacy Clause sets forth a deceptively simple rule for resolving the conflict: the laws of the United States shall be the supreme law of the land.[66] In practice, the determination of when and to what extent federal enactments supersede state law is often a painstaking process involving the application of what the courts have labeled the doctrine of federal preemption. Generally speaking, whether state law is deemed preempted is largely a matter of statutory construction,[67] the role of the court being to ascertain congressional intent as to whether the states should be allowed to regulate in a field also occupied by the federal government. It is often said that where a federal statute conflicts with a state law concerning a subject historically within the scope of the police power, the federal statute will not be held to have superseded the state's police power "unless that was the clear and manifest purpose of Congress."[68]

Congressional intent to preempt may be manifested either explicitly or implicitly. The simplest case is where Congress has expressly indicated on the face of the statute or in the legislative history that states are to be precluded from regulating over the same subject matter.[69] Similarly, preemption will be found where a state law actually conflicts with federal law so that compliance with both regulations is physically impossible,[70] or where the state law "stands as an obstacle to the accomplishment and executive of the full purposes and objectives of Congress."[71]

Even in the absence of actual conflict or of an express declaration of preemptory intent, the courts will infer such an intent under certain circumstances. Thus, where the scheme of federal regulation is so pervasive as to support a reasonable conclusion that Congress left no room for state action, preemption will be implied.[72] Likewise, state laws will give way where the congressional enactment touches upon a "field in which the federal interest is so dominant that the federal system will be assumed to preclude enforcement of state laws on the same subject."[73] Finally, a congressional purpose to foreclose state regulation may be divined where "the object sought to be obtained by the federal law and the obligations imposed by it reveal" such an intent on the part of Congress.[74]

In the field of nuclear energy regulation, the cornerstone of the preemption analysis is section 274 of the Atomic Energy Act of 1954, which, as we have seen, purports to delimit the correlative powers of the states and the federal government with respect to the regulation of radiation hazards connected with the use of the three classes of nuclear material: source, special nuclear, and by-product.[75] Significantly, section 274(b) makes specific mention of nuclear waste disposal and prohibits the NRC from delegating any authority in this area to the states where the commission determines that such material represents a special hazard.[76] As noted above, the NRC has retained such authority with respect to HLW, but has permitted agreement states to license and regulate commercial LLW burial grounds.[77] The clear implication of the statutory language is that absent a relinquishment of authority by the NRC, the states would have no power to regulate radioactive wastes.[78]

From all indications, it would appear, furthermore, that the AEC and the Congress devoted substantial attention to the preemption question, opting in favor of a scheme that would vest the federal government with the exclusive power over the regulation of radiation levels due to nuclear activities, subject to delegation to the states by agreement, but which preserved the state's police power in peripheral areas which do not directly involve radiological safety.[79]

The record is quite clear that Congress intended to avoid any possibility of dual or concurrent state and federal regulation. So long as the federal government did not relinquish its regulatory authority, the states were to be powerless to act in the area of radiation safety. Thus, according to a report of the Joint Committee on Atomic Energy, section 274 was not intended

> to leave any room for the exercise of dual or concurrent jurisdiction by States to control radiation hazards by regulating byproduct, source, or special nuclear materials. The intent is to have the material regulated and licensed either by the Commission, or by State and local governments, but not by both.[80]

During the course of the hearings on the AEC bill which was ultimately to become section 274, members of the Joint Committee and AEC staff personnel debated at length the issue of preemption. At one point, committee members proposed an amendment to the language of what is now section 274(k) which provided in part:

> It is the intention of this Act that State laws and regulations concerning the control of radiation hazards from byproduct, source, and special nuclear materials shall not be applicable except pursuant to an agreement entered into with the Commission pursuant to subsection b;[81]

The AEC took strong exception to the need for such additional language and urged its deletion, stating in a letter from its general manager to the chairman of the Joint Committee:

> At the hearing, we recommended the omission of the sentence. In making this recommendation, we did not intend to change the substantive effect of the bill because we believe that under this bill, with or without the sentence, the Federal Government will clearly have "preempted" the regulation and control of radiation hazards from source, byproduct, and special nuclear materials. In either event, preemption will end in any State only upon the effective date of an agreement between the state and the Commission under subsection b, and only to the extent provided in the agreement.
>   In suggesting the elimination of the sentence, we did not intend to leave any room for the

exercise of concurrent jurisdiction by the States to control radiation hazards from those materials. Our sole purpose was to leave room for the courts to determine the applicability of particular State laws and regulations dealing with matters on the fringe of the preempted area in the light of all the provisions and purposes of the Atomic Energy Act, rather than in the light of a single sentence.

For example, in the absence of the sentence, the courts might have greater latitude in sustaining certain types of zoning requirements which have purposes other than control of radiation hazards, even though such requirements might have an incidental effect upon the use of source, byproduct, and special nuclear materials licenses by the Commission.[82]

In deference to the AEC, the Joint Committee eliminated its proposed language and reported out section 274(k) in its present form.[83]

Since the enactment of section 274, several federal and state courts have confronted the issue of federal preemption under the 1954 act.[84] With few exceptions, the judicial view of section 274 has been and is that in the absence of an agreement with the NRC, the states are prohibited from legislating in the field of radiation hazards even where the measures in question are deemed necessary to protect the health, safety, and welfare of the public at large.

The seminal and landmark case on this question of preemption is the Eighth Circuit Court of Appeals' decision in *Northern States Power Co.* v. *Minnesota.*[85] In that case, the state of Minnesota attempted to regulate the release of radioactive emissions from a privately owned nuclear power plant already licensed by the AEC under the 1954 act. The conditions imposed by the Minnesota Pollution Control Agency were substantially more restrictive than those imposed under federal law.

The court determined initially, based in part on a concession by the parties, that the Atomic Energy Act did not expressly grant the federal government sole and exclusive authority to regulate radioactive emissions from nuclear power plants.[86] The court went on to conclude, however, based on its review of the statutory language, that the pervasiveness of the federal scheme and history of the 1946 and 1954 acts and the 1959 amendment that Congress had intended to preempt the field. In response to the state's argument that section 274 sanctions the exercise of concurrent federal and state jurisdiction over the construction and operation of nuclear power plants, the court stated emphatically:

We cannot agree with Minnesota's position that dual control over atomic power plants and the level of effluents discharged therefrom is permissible under the Act. While the 1959 amendment does not use the terms "exclusive" or "sole"in describing existing regulatory responsibilities of the Commission, we think it abundantly clear that the whole tone of the statutory language alone, demonstrates Congressional recognition that the AEC at that time possessed the sole authority to regulate radiation hazards associated with byproduct, source, and special nuclear materials and with production and utilization facilities.[87]

Although the precise holding of *Northern States* has now been superseded by subsequent legislative enactment in the environmental protection area,[88] the Eighth Circuit's conclusions with regard to preemption remain intact and have received the apparent blessings of the United States Supreme Court in the recent case of *Train* v. *Colorado Pub. Interest Research Group, Inc.*[89] The issue in the *Colorado PIRG* decision was whether the Environmental Protection Agency had authority to regulate the discharge of nuclear waste materials into the nation's waterways under the Federal Water Pollution Control Act, as amended.[90]

In reviewing the legislative history of the FWPCA, the Court noted that Congress was concerned over whether the FWPCA would impinge upon the authority of the AEC vis-a-vis the states to regulate by-product materials, that is, radioactive wastes. In discussing the significance of congressional awareness of the *Northern States* decision, the Court observed:

> In *[Northern States],* which was subsequently affirmed summarily by this Court . . . [citation omitted], the Eighth Circuit held that the AEA created a pervasive regulatory scheme, vesting exclusive authority to regulate the discharge of radioactive effluents from nuclear power plants in the AEC, and pre-empting the States from regulating such discharges. The absence of any room for a state role under the AEA in setting limits on radioactive discharges from nuclear power plants stands in sharp contrast to the scheme created by the FWPCA, which envisions the development of state permit programs . . . and allows the States to adopt effluent limitations more stringent than those required or established under the FWPCA. . . . Senator Muskie's specific assurance to Senator Pastore that the FWPCA would not affect existing law as interpreted in *Northern States* can only be viewed, we think, as an indication that *the exclusive regulatory scheme created by the AEA for source, byproduct, and special nuclear materials was to remain unaltered.* [Emphasis added.][91]

More recently, at least two federal district courts and several state courts[92] have adopted the reasoning set forth in *Northern States* to strike down state legislation that encroached upon the federal regulatory domain. Of particular interest is *Pacific Legal Foundation* v. *State Energy Resources Conservation & Development Commission,*[93] which involved a successful challenge to a California statute which preconditioned state licensing of nuclear power plants upon the issuance of certificate by a state commission to the effect that the federal government has approved a technology for the disposal of high-level nuclear wastes.[94] In striking down the certification requirement, the district court in *Pacific Legal Foundation,* relying heavily on *Northern States,* declared:

> the court finds California Public Resources Code section 25524.2 preempted both because Congress *has impliedly foreclosed state legislation on the subject of nuclear waste* and, alternatively, because the statute stands as an obstacle to the purposes and objectives of Congress as stated in the Atomic Energy Act of 1954, as amended.[95] [Emphasis added.]

The court went on to explain:

> Section 2021(c) [AEA section 274(c)] provides that the NRC shall retain authority and responsibility with respect to the regulation of the construction and operation of nuclear

power plants and with respect to the regulation of nuclear waste disposal. In the exercise of its discretion, the NRC has decided not to require the existence of a technology for the permanent disposal of nuclear waste as a condition precedent for the construction and operation of nuclear reactors. The NRC's decision in this regard falls within the preempted sphere because it relates to, touches upon and involves the regulation of radiation hazard pertaining to the construction and operation of nuclear power plants and to nuclear waste disposal. California has decided otherwise, decreeing that no nuclear power plant may be constructed in the State of California unless there exists a demonstrated technology for the disposal of nuclear waste. The court finds that the question of whether nuclear power plants may be constructed and operated in the absence of a demonstrated technology for the permanent disposal of nuclear waste is exclusively reserved to the NRC by section 2021(c) and that state regulation on this subject is displaced. Accordingly, the court holds California Public Resources Code section 25524.2 impliedly preempted.[96]

Alternatively, the court reasoned that California's waste disposal certification requirement would, if allowed to stand, frustrate one of the principal purposes behind the Atomic Energy Act, namely, to encourage the development and utilization of nuclear energy.

Although the Atomic Energy Act certainly leaves room for the states to regulate on the subject of nuclear energy within the confines of section 2021(k) and 2021(b), the power to regulate is not necessarily the power to prohibit. There seems to be little point in enacting an Atomic Energy Act and establishing a federal agency to promulgate extensive and pervasive regulations on the subject of construction and operation of nuclear reactors and the disposal of nuclear waste if it is within the prerogative of the states to outlaw the use of atomic energy within their borders.[97]

Also of interest is the recent decision of the United States District Court for the Southern District of New York in *United States* v. *City of New York,*[98] where the United States and the trustees of Columbia University brought a challenge on preemption grounds to a New York City ordinance requiring a city permit before operation of a nuclear reactor could be commenced. After an exhaustive review of the legislative history of section 274 and a lengthy discussion of the *Northern States* decision, the court ruled that the city ordinance unconstitutionally intruded upon a field committed exclusively to the federal government. Worthy of mention is the court's flat rejection of the city's contention that section 274 allows states leeway in promulgating regulations governing the siting of nuclear reactors even where such regulations are not based on radiological safety criteria. In the court's view, Congress prohibited any and all regulation by states of radiation hazards associated with reactor operation.[99]

Northern States and its recent progeny would seem to close the door on the assertion of state authority to prohibit or severely impede the siting and establishment of nuclear waste repositories, except to the extent such powers are delegated by the NRC pursuant to a section 274(b) turnover agreement. It is true that the decided cases were concerned primarily with state regulation of nuclear power plants and did not specifically deal with nuclear waste disposal bans such as have been enacted by several states. Nevertheless, if state intrusions upon NRC authority in the power plant area are precluded, the case against state regulation of waste repositories would be even more compelling. This is because, unlike decisions respecting the construction of power facilities which involve important nonradiological considerations, the principal, if not the sole, motivation for regulating waste repositories would be to protect the public against radiological harm. This, as we have seen, is an area into which the states simply may not intrude except by contract with the NRC.

As previously noted, section 274(k) of the Atomic Energy Act of 1954, as amended, expressly authorizes state regulation of NRC licensed activities for purposes other than protection against radiation hazards.[100] The now defunct Joint Committee on Atomic Energy explained in its report accompanying the legislation eventually enacted as section 274:

> [S]ubsection K is intended to make it clear that the bill does not impair the state[s'] authority to regulate activites of AEC licensees for the manifold health, safety and economic purposes other than radiation protection.[101]

Thus, it has been held that the Atomic Energy Act would not prohibit state agencies from examining such nonradiological hazards as the environmental effect of steam, fog, and winter icing engendered by a nuclear power plant cooling pond in making a determination whether to approve the siting of the plant.[102]

Traditionally, states utilized their police powers to regulate a wide variety of "local" matters which could impact substantially on the siting, construction, and operation of nuclear facilities, including radioactive waste repositories, but which do not necessarily involve the regulation of radiological hazards. Examples of such possible nonradiological controls include zoning legislation, air and water pollution regulations, fish and wildlife protection measures, soil and water conservation ordinances, and legislation for the licensing of power generating facilities.[103]

Conceivably, a state or locality could invoke its long-recognized authority to impose land-use controls so as to bar completely the location of a nuclear waste facility within its territorial jurisdiction. Theoretically, under section 274(k), such a use of the zoning power might avoid a federal preemption challenge if it is not aimed at protecting the public from the hazards of radiation, a subject which, as we have seen, falls within the exclusive domain of the national government.[104] Moreover, the Supreme Court has said that, where Congress legislates in a field such as land-use planning which the states have traditionally occupied, there is a presumption that the historical police powers of the states are not to be ousted by a federal statute "unless that was the clear and manifest purpose of Congress."[105] On closer scrutiny, however, it should be apparent that as a practical (and legal) matter, the invocation of the zoning power to exclude nuclear waste storage of disposal facilities would almost always be prompted by radiological considerations. In the first place, most, if not all, of the present state antinuclear waste legislation is the direct outgrowth of a widespread public concern over the nature and scope of the potential radiological threat that temporary or permanent radioactive waste storage and disposal installations might pose.[106] Public attention has not been focused on the aesthetics of waste repositories, for example; rather, the debate to date has concentrated almost exclusively on the radiation dangers associated with nuclear waste disposal. Consequently, any subsequent attempt by the states to ban the disposal facilities through the enactment of a facially nonradiologically oriented zoning regulation would likely be seen by the courts as a thinly veiled attempt to circumvent the rather broad restrictions on state regulatory authority under the Atomic Energy Act.

Were states and local governments totally free to employ the zoning power to prohibit the establishment of nuclear waste facilities on nonfederal land,[107] it is quite plausible that the unchecked proliferation of such restrictions would seriously undermine, if not completely destroy, the nation's nuclear energy program. Without passing judgment on the ultimate desirability of nuclear power as a matter of public policy, it is unmistakable that under the current legislative framework, the Congress has committed the federal government to the continued development of nuclear energy resources to the extent that such development will not endanger either the public health and safety or the environment.[108] Inasmuch as the construction of temporary and permanent installations for the storage and disposal of radioactive waste is indispensable to the nation's nuclear energy program, any serious interference by the states with the establishment of a national waste management program under the guise of the zoning power would likely stand "as an obstacle to the accomplishment and execution of the full purposes and objectives of Congress" in enacting the Atomic Energy Act and subsequent nuclear-related legislation.[109] State action which has the effect of discouraging conduct or programs that the federal legislation seeks to promote violates the Supremacy Clause and is therefore void.[110]

Left uncontrolled, state regulation (through any medium) of waste disposal could conceivably frustrate any attempts to formulate a uniform and effective national waste management policy. As of this writing, nearly half the states have passed legislation prohibiting or reserving the right to veto nuclear waste installations within their borders. If current trends continue and the remaining jurisdictions adopt similar restrictions, whether through zoning legislation or otherwise, the federal government's ability to site and erect sorely needed waste repositories would be severely curtailed. The possibility of such interference by the states with the implementation of federal policy is sufficient for purposes of the Supremacy Clause to trigger the doctrine of preemption and render the potentially conflicting state legislation invalid.

In sum, whatever form state anti-waste-disposal legislation takes, be it a flat prohibition or a more indirect exclusion through zoning or environmental controls, the prospect that it will withstand judicial scrutiny on preemption grounds is rather bleak, at lest under the existing statutory regime.

Before leaving the subject of preemption, a brief look at the problem of state regulation of nuclear waste transport may be in order. Under the Atomic Energy Act of 1954, the NRC is vested with full authority to regulate the transportation of source, special nuclear, and by-product materials.[111] However, the NRC has specifically exempted common and contract carriers from its jurisdiction.[112] By agreement with the United States Department of Transportation (DOT),[113] NRC licenses and regulates the shipment and receipt of radioactive materials and, in addition, sets safety standards for the packaging of certain types of nuclear material. For its part, DOT sets standards for packaging and labeling of materials not within NRC's purview. In addition, DOT develops and enforces safety standards governing the mechanical conditions of carrier equipment and qualifications of carrier personnel, carrier loading, unloading, handling, and storage of radioactive material, and any special transport controls to be provided during carriage.[114] DOT is currently in the process of developing regulations for the routing of highway shipments of nuclear wastes.[115]

Highway traffic control, maintenance, and regulation of intrastate motor carriers have long been the subject of state regulation.[116] Nevertheless, transportation is an essential link in any national nuclear waste program.

To the extent that states and local governments purport to restrict the initial transfer and receipt of nuclear materials, such action would appear to be preempted under the Atomic Energy Act of 1954 as implemented by the NRC. State regulation of nuclear waste motor carriers poses a somewhat different problem inasmuch as NRC had elected to occupy this area under the 1954 Act.

The regulation of carriers, as we have noted, is within the domain of DOT, which derives its authority over radioactive materials principally from the Hazardous Materials Transportation Act,[117] as amended. Under this statute, DOT is authorized to issue regulations "for the safe transportation in commerce of hazardous materials."[118] Such regulations may deal with any safety aspect of the transportation of hazardous materials including packing, handling, labeling, and routing.[119] Pursuant to this authority, DOT has promulgated a battery of regulations dealing extensively with the transport of radioactive substances.[120]

Unlike the Atomic Energy Act of 1954, the Hazardous Materials Transportation Act provides for a significant degree of state regulation in the field. The Hazardous Materials Act permits state and local governments to impose restrictions on the transport of hazardous materials, provided such restrictions are authorized by the Secretary of Transportation in accordance with procedures set forth in the regulations[121] Two criteria are established for the non-preemption of state law. First, the state requirement must provide "an equal or greater level of protection to the public than is afforded" by federal law.[122] Second, the state regulations must not "unreasonably burden commerce."[123]

Given the elaborately detailed provisions regarding preemption set forth in the federal legislation and in the administrative regulations thereunder, there can be little if any doubt that Congress intended to occupy the field of hazardous material transport. This is evidenced by the requirement for advance federal approval of state and local restrictions. The implication of the statute and the regulations is that, absent approval by DOT, a conflicting action within the purview of the statute would be preempted.[124]

The regulations, do, however, provide that preemption rulings issued by DOT constitute

> an administrative determination as to whether a particular requirement of a State or local subdivision is inconsistent with the Act or the regulations issued under the Act. The fact that a ruling has not been issued under this section with respect to a particular requirement of a State or political subdivision carries no implication as to the consistency or inconsistency of that requirement with the Act or any regulations issued under the Act.[125]

State statutes or regulations which ban the transport of nuclear wastes altogether would appear to contravene federal law in that they would constitute unreasonable burdens on commerce. This topic is discussed in greater detail in the following section. With respect to those state statutes which restrict nuclear waste transport short of a total prohibition, the question of preemption is more difficult and must, in the final analysis, be resolved on a case-by-case basis.

Commerce Clause

At least three states have enacted legislation which would bar the importation of radioactive wastes for purposes of disposal within the jurisdiction.[126] One state, Louisiana, has gone so far as to close its borders to any high-level wastes being transported for burial in the state or elsewhere.[127] Aside from the probable infirmity of such statutes under the preemption doctrine, [128] they would also appear to be subject to attack on the grounds that they constitute an unconstitutional extension of state regulatory authority over interstate commerce.[129]

Article I, Section 8 of the United States Constitution specifically grants to Congress the power to regulate interstate and foreign commerce. Despite the existence of this power, the Supreme Court has repeatedly held that, in the absence of federal regulation covering a particular aspect of interstate commerce, the states are free to legislate on subjects of a local character, provided they are within the restraints imposed by the Commerce Clause.[130] State regulation of interstate commerce will be sustained where the legislation does

not patently discriminate against interstate trade, where the legislation promotes a legitimate local interest, and where its effects on interstate commerce
are not unduly unburdensome. As the Court held in *Pike* v. *Bruce Church,
Inc.*:

> Where the statute regulates evenhandedly to effectuate a legitimate local public interest, and
> its effects on interstate commerce are only incidental, it will be upheld unless the burden
> imposed on such commerce is clearly excessive in relation to the putative local benefits . . . If
> a legitimate local purpose is found then the question becomes one of degree. And the extent
> of the burden that will be tolerated will of course depend on the nature of the local interest
> involved, and on whether it could be promoted as well with a lesser impact on interstate
> activities.[131]

Where, however, the state legislation patently discriminates against interstate commerce so as to promote essentially protectionist objectives, the Court
is reluctant to engage in the type of flexible balancing analysis described in
Pike. Rather, in such cases, the Court appears to apply "a virtually per se rule
of invalidity."[132]

The recent case of *Philadelphia* v. *New Jersey*[133] is of particular relevance in
light of its factual similarity to the question presumably under discussion. At
issue in *Philadelphia* was a New Jersey statute which prohibited the importation of most solid or liquid waste which originated or was collected outside the
territorial limits of the state. The express purpose of the legislation was to
protect the New Jersey environment and the public health and safety by
reducing the quantity of waste imported into the state.[134]

Despite these and other salutary reasons for New Jersey's prohibition on
imported waste, the Supreme Court in a 7-2 decision did not hesitate to strike
down the statute as violative of the Commerce Clause.

> The New Jersey law at issue in this case falls squarely within the area that the Commerce
> Clause puts off-limits to state regulation. On its face, it imposes on out-of-state commercial
> interests the full burden of conserving the State's remaining landfill space.[135]

The state had argued that the New Jersey statute could be sustained on the
authority of the quarantine law cases, such as *Absell* v. *Kansas*,[136] which
upheld state laws banning the importation of articles such as diseased livestock
even though they appeared to single out interstate commerce for special
treatment.

Rejecting the proffered parallel with the quarantine cases, the Court, speaking through Justice Stewart, declared:

> The New Jersey statute is not such a quarantine law. There has been no claim here that the
> very movement of waste into or through New Jersey endangers health, or that waste must be
> disposed of as soon and as close to its point of generation as possible. The harms caused by
> waste are said to arise after its disposal on landfill sites, and at that point, as New Jersey

concedes, there is no basis to distinguish out-of-state waste from domestic waste. If one is inherently harmful, so is the other. Yet New Jersey has banned the former while leaving its landfill sites open to the latter. The New Jersey law blocks the importation of waste in an obvious effort to saddle those outside the State with the entire burden of slowing the flow of refuse into New Jersey's remaining landfill sites. That legislative effort is clearly impermissable under the Commerce Clause of the Constitution.

Today, cities in Pennsylvania and New York find it expedient or necessary to send their waste into New Jersey for disposal, and New Jersey claims the right to close its borders to such traffic. Tomorrow, cities in New Jersey may find it expedient or necessary to send their waste into Pennsylvania or New York for disposal, and those States might then claim the right to close their borders. The Commerce Clause will protect New Jersey in the future, just as it protects her neighbors now, from efforts by one State to isolate itself in the stream of interstate commerce from a problem shared by all.[137]

Current state bans on nuclear waste importation are similar in many respects to the New Jersey solid waste statute invalidated in *Philadelphia*. Like the New Jersey statute, such bans apply only to incoming commerce and would not appear to ban disposal of locally generated wastes. Like the New Jersey statute, nuclear waste bans have the effect of isolating the state from the rest of the economy. Like the New Jersey statute, blanket prohibitions on radioactive wastes were enacted to protect the public from grave harm.

On the other hand, anti-nuclear-waste statutes bear some similarity to the quarantine laws (of which the Court appeared to approve), since it may be argued that the mere transportation of radioactive waste into a state might endanger the public safety. The danger does not arise according to this argument after the waste is stored or disposed, as was the case with New Jersey's solid wastes.

While the result is by no means certain, a state ban on foreign nuclear wastes without a concomitant prohibition against the disposal of domestic wastes would appear to have the effect of "saddling those outside the State with the entire burden" of nuclear waste. It is precisely this type of parochialism which, in the Supreme Court's view, is most offensive to the objectives of the Commerce Clause and the myriad judicial decisions thereunder. Under these circumstances, the validity of statues such as those of Montana, Minnesota, and Louisiana is doubtful at best.[138] Moreover, even if a per se rule of invalidity is not applied to such statutes and, instead, a flexible balancing approach is invoked, a substantial question remains as to whether the states' legitimate local interest in safeguarding the public against radiation hazards due to nuclear waste transport might have been effectuated by less drastic means.[139] If the state regulation could have been more narrowly tailored (for example, by means of a registration and inspection requirement), the balance would clearly tip against the enforceability of the state statute.

CONCLUSION

The present controversy over nuclear power will continue to focus public attention increasingly on the enormously complicated problem of nuclear waste management. Certain laudable attempts by state governments to fill the breach created by years of regulatory neglect and incompetence at the federal level by means of siting moratoria and embargos, given the existing legal regime, are likely unconstitutional. To the extent states purport to regulate the disposition of HLW and LLW on federal reservations, the doctrine of intergovernmental immunity would preclude enforceability. More importantly, it would appear that any state intrusion into the field of nuclear waste management would at this time be preempted by the Atomic Energy Act of 1954, as amended, and the regulations issued pursuant thereto. Finally, legislative barriers imposed by states to the importation of out-of-state nuclear wastes likely run afoul of the Commerce Clause.

Not withstanding the unfavorable legal climate that now prevails, state participation in the licensing and regulation of radioactive waste repositories is an idea whose time has certainly come. Local protest has figured heavily in the abandonment of two federal waste disposal projects in the past[140] and many have forestalled a third indefinitely.[141] If states are to achieve the degree of meaningful participation for which they now clamor and for which the federal government appears willing to allow, a substantial readjustment in the allocation of power to federal and state governments within the existing nuclear energy legislative and regulatory scheme should be high on the list of congressional priorities in the coming session.[142]

NOTES

1. Statement of Commissioner Richard T. Kennedy, United States Nuclear Regulatory Commission to the Utility Regulatory Conference, Washington, D.C., October 5, 1978. See also Interagency Review Group on Nuclear Waste Management, *Report to the President,* March 1979, T1D-29442 (hereafter referred to as IRG, *Report*). Dr. Harvey Brooks has stated: "No single aspect of nuclear power has excited so persistent a public concern as has radioactive waste management. . . . I would predict that, should nuclear energy ultimately prove to be socially acceptable, it will be primarily because of the public perception of the waste disposal problem." 195 *Science* 661 (1977).

According to data set out in the IRG report, there presently exist in the United States approximately 9.4 million cubic feet of high-level waste, 1200 kilograms of transuranic waste, 2300 metric tons of spent fuel from commercial reactors, 66.6 million cubic feet of buried low-level wastes, and 140 million tons of uranium mill tailings. IRG, *Report,* Appendix D.

2. See Daniel S. Metlay, *"History and Interpretation of Radioactive Waste Management in the United States,"* in U.S. NRC Office of Nuclear Material Safety and Safeguards, *Essays on Issues Relevant to the Regulation of Radioactive Waste Management,* NUREG-0412 (May 1978). See also *Nuclear Waste Management: Hearings Before the Subcomm. on Energy, Nuclear Proliferation and Federal Services of the Sen. Comm. on Gov'tal Affairs,* 95th Cong., 2d Sess. 4 (1978) (statement of Sen. Javits). For a brief but good discussion of the history of federal waste management policy

from 1959-1977, see *Natural Resources Defense Council, Inc.* v. *U.S. Nuclear Reg. Comm'n,* 582 F.2d 166, 170 *et seq.* (2d Cir. 1978).

3. Montana (MONT. REV. CODES ANN. Section 75-3-302), Alabama (H-176, approved 5/14/79), and North Dakota (S-2168, approved 3/8/79), for example, flatly prohibit the disposal within their territorial jurisdictions of radioactive waste generated in other states, while Vermont bans the construction or establishment of high-level radioactive waste facilities without the prior approval of the state legislature. VT. STAT. ANN. Tit. 10, Section 6501. Similar statutes are in force in Colorado (S-335, approved 6/15/79), Connecticut (H-5097, approved 6/18/79), Louisiana [LA. REV. STAT. ANN. Section 1071 (West)], Maine (H-799, approved 6/22/79), Maryland [MD. ANN. CODE ART. 49, Section 689B (1957, 1971 Repl. Vol. & 1978 Cum. Supp.)], Michigan [MICH. STAT. ANN. Section 14.528 (351)], Minnesota [MINN. STAT. ANN. Section 116C.71 (West)], New Hampshire (H-91, approved 6/23/79), Oregon (OR. REV. STAT. Section 459,625), and South Dakota (1977 S.D. Sess. Laws, Ch. 283). Legislation recently enacted in New York and New Mexico requires that the state consult and concur prior to the siting of a permanent waste repository (NY: A-7363, A-3197B, approved 7/11/79) (H-106, 360, 500, 527, approved 4/6/79). The legislatures of Delaware and Hawaii have passed formal resolutions memorializing Congress to eliminate their respective states from consideration as sites for future waste repositories. House Resolution, HR-124 (Del. April 14, 1978); Radioactive Waste Disposal in the Pacific Ocean, SR-68 (Haw. April 5, 1976).

The construction of nuclear power plants is banned by statute in California [CAL. PUB. RES. CODE Section 25524.2 (West)], Maine (H-1338, June 22, 1977), and Connecticut (H-5096, approved 6/18/79), and by administrative fiat in Wisconsin [Docket No. 05-EP-1 (Wis. Pub. Serv. Comm., August 17, 1978)] until the respective state agencies certify that a demonstrated technology for the disposal of high-level nuclear wastes exists.

In April 1979, the governor of South Carolina notified the U.S. Nuclear Regulatory Commission that his state would not accept low-level wastes from the crippled Three Mile Island nuclear power plant or from any nuclear power facility shut down unexpectedly. (Letter from Hon. Richard W. Riley, Gov. of S.C., to Joseph M. Hendrie, Chmn., NRC, dated April 19, 1979.) In October 1979, the governors of Nevada and Washington ordered the commercial waste burial grounds in their states to cease receiving wastes, due to potential radiological hazards engendered by inadequate enforcement of federal waste transportation standards, as well as the improper burial of low-level wastes. 20 *Nucleonics Week* no. 41 (Oct. 11, 1979); *Journal of Commerce* vol. 3 (Oct. 25, 1979), p.1.

Several states have enacted legislation severely restricting the transport of radioactive wastes within their jurisdictions. For example, Connecticut [CONN. GEN. STAT. ANN. Section 19-409d (West)]; Louisiana [LA. REV. STAT. ANN. Section 1072 (West)]; Minnesota (MINN. STAT. ANN. Section 1116C.73).

See generally, U.S. NRC, Office of State Programs, *Means for Improving State Participation in the Siting, Licensing and Development of Federal Nuclear Waste Facilities,* NUREG-0539, App. D (March 1979) ("NRC-State Participation"); U.S. N.R.C., Office of State Programs, 5 *Information Report on State Legislation* (Nos. 7-15) (1979).

4. U.S. NRC, *Means for Improving State Participation,* p.1.

5. See IRG, *Report,* p.88 *et seq.;* U.S. N.R.C., Office of Nuclear Material Safety and Safeguards, *NRC Task Force Report on Review of the Federal/State Program for Regulation of Commercial Low-Level Radioactive Waste Burial Grounds,* NUREG-0217, (1977), pp. 12-13. See also "Special Message of President Carter to Congress on the Environment," 125 *Congressional Record* 7132 (1979).

6. The Atomic Energy Act of 1954, 68 Stat. 919, *as amended,* 42. U.S.C.A. Sections 2011-2296 (1979).

7. Energy Reorganization Act of 1974, Pub. L. No. 93-438, 88 Stat. 1233 (1974).

8. Department of Energy Organization Act, Pub. L. No. 95-91, 91 Stat. 565 (1977).

9. Uranium Mill Tailings Radiation Control Act of 1978, Pub. L. 95-604, 92 Stat. 3021.

10. The Atomic Energy Act of 1946, ch. 724, 60 Stat. 755.

11. Idem; J. R. Newman, "The Atomic Energy Industry: An Experiment in Hybridization," 60 *Yale L.J.* 1263 (1951); *Northern States Power Co.* v. *Minnesota,* 447 F.2d 1143, 1143 (8th Cir. 1971), *aff'd mem.,* 405 U.S. 1035 (1972).

12. NRC control over the three classes of nuclear material is all-encompassing. With respect to the licensing of by-product material, for example, the 1954 Act currently provides: "No person may transfer or receive in interstate commerce, manufacture, produce, transfer, acquire, own, possess, import, or export any byproduct material, except to the extent authorized by this section, section 2112 or section 2114 of this title. 42 U.S.C.A. Section 2111 (1979)." Similar prohibitions govern the use of special nuclear [42 U.S.C. Section 2073(a)(1976)] and source material [42 U.S.C. Section 2092 (1976)]. Definitions of the three categories of regulated material are set out immediately below.

"Source material" [42 U.S.C. Sections 2014(z), 2091 (1976)] is defined to include uranium, thorium, and such other materials as the Nuclear Regulatory Commission determines are essential to the production of special nuclear material. Also included within this definition are certain ores containing uranium, thorium, or NRC-designated materials. Source materials represents the raw material from which special nuclear material is derived. 10 C.F.R. Section 40.4(h) (1979).

"Special nuclear material" [42 U.S.C. Section 2014(aa), 2077 (1976)] is defined as plutonium, uranium 233, uranium enriched in the isotope 233 or 235, and any other material designated by the commission to be special nuclear material; or any material enriched by any of the foregoing. 10 C.F.R. Section 70.4(m) (1979).

"By-product material" [42 U.S.C.A. Sections 2014(e)(1), 2111 (Supp. 1979)] is defined as any radioactive material, except special nuclear material, yielded or made radioactive by exposure to the radiation incident to the process of producing or utilizing special nuclear material. 10 C.F.R. Section 30.4(d) (1979). As a consequence of the enactment of the 1978 Uranium Mill Tailings Radiation Control Act, by-product material now also includes wastes produced by the extraction or concentration of uranium or thorium from any ore processed primarily for its content of source material. 42 U.S.C.A. Section 2014(e)(2) (Supp. 1979).

13. See *City of New Britain* v. *U.S. Atomic Energy Comm'n,* 308 F.2d 648, 649 (D.C. Cir. 1962) and *Harris County* v. *United States,* 292 F.2d 370, 371 (5th Cir. 1961), both holding that "by-product material" encompasses radioactive wastes within the authority of the Commission to control.

14. 42 U.S.C. Sections 2073 (e)(7), 2099, 2133(b), (d), 2134(a)(d) (1976).

15. Act of September 23, 1959, Pub. L. No. 86-373, 73 Stat. 688, *as amended* by 1970 Reorg. Plan No. 3, 84 Stat. 2086, 5 U.S.C. App. (1976).

16. 42 U.S.C. Section 2021(c) (4) (1976). As a result, companies operating low-level waste disposal sites in agreement states must obtain licenses from NRC for special nuclear material as well as from state authorities. See *Mississippi Power & Light Co.* v. *U.S. Nuclear Reg. Comm'n,* 601 F.2d 223, 233 (5th Cir. 1979).

17. U.S. NRC, Office of State Programs, *Final Task Force Report on the Agreement States Program,* NUREG-0388, 1977, p. A-53. Rhode Island became the latest state to participate in the program on October 23, 1979. U.S. NRC *Proposed Agreement Between the State of Rhode Island and the NRC,* SECY 79-503 (Oct. 23, 1979)—hereafter referred to as NRC, *Agreement States Report.* For a detailed discussion of the legislative history of section 274, see NRC, *Agreement States Report,* p. A-1 et seq.

18. 42 U.S.C. Section 2021(k) (1976). See generally Arthur W. Murphy & D. Bruce La Pierre, "Nuclear 'Moratorium' Legislation in the States and the Supremacy Clause: A Case of Express Preemption," 76 *Colum. L. Rev.* 392, 450-454 (1976).

19. Pub. L. No. 93-438, Section 202(3)-(4), 88 Stat. 1244, 42 U.S.C. Section 5842(3)-(4) (1976). See also S. Rep. No. 93-980, 93d Cong., 2d Sess. 59, [1974] *U.S. Code Cong. & Ad. News* 5470, 5521; H. Rep. No. 93-1445, 93d Cong. 2d Sess., [1974] *U.S. Code Cong. & Ad. News* 5546-47 (Conference Report). And see *Natural Resources Defense Council* v. *U.S. Nuclear Reg. Comm'n,*

606 F.2d 1261, 1266-1268 (D.C. Cir. 1979).

20. Pub. L. No. 93-438, Section 203(b)(2)(B), 88 Stat. 1244, 42 U.S.C. Section 5843(b)(2) (B) (1976).

21. Pub. L. No. 93-438, Section 204(b) (2) (B), 88 Stat. 1245, 42 U.S.C. Section 5844(b)(2)(B) (1976).

22. Pub. L. No. 95-91, 91 Stat. 565, 42 U.S.C. Section 7101 *et seq.* (Supp. I, 1977).

23. Pub. L. No. 95-91, Section 203(8), 91 Stat. 571, 42 U.S.C. Section 7133(a)(8) (Supp. I, 1977).

24. *Idem.*

25. Pub. L. No. 95-691, Section 14(a), 92 Stat. 2953, 42 U.S.C.A. Section 2021a(a) (Supp. 1979). See also *H. Rep. No. 95-1800,* 95th Cong., 2d Sess. 17, reprinted in [1978] *U.S. Code Cong. & Ad. News* 7310.

26. Pub. L. No. 95-601, Section 14(b), 92 Stat. 2953, 42 U.S.C.A. Section 2021a(b) (Supp. 1979).

27. Uranium Mill Tailings Radiation Control Act of 1978, Pub. L. 95-604, 92 Stat. 3021.

28. *Idem.,* Section 204(e)(1), 42 U.S.C.A. Section 2021(o) (Supp. 1979).

29. *Idem.,* Section 202, 42 U.S.C.A. Section 2113 (Supp. 1979).

30. For example, S. 1360, 96th Cong., 1st Sess., 125 *Cong. Rec.* 7882 (1979) (state concurrence through agreement); S.701, 96th Cong., 1st Sess., 125 *Cong. Rec.* 3059 (1979) (state veto by legislature or referendum); S. 1521, 96th Cong., 1st Sess., 125 *Cong. Rec.* 9521 (1979) (meaningful state participation); S. 1443, 96th Cong., 1st Sess., 125 *Cong. Rec.* 8710 (1979) (state veto); S. 594, 96th Cong., 1st Sess., 125 *Cong. Rec.* 2407 (1979) (state participation and concurrence).

31. See, e.g., U.S. NRC, Office of Nuclear Material Safety and Safeguards, *A Classification for Radioactive Waste Disposal - What Waste Goes Where?,* NUREG-0456 (1978).

32. 10 C.F.R. Section 50, App. F(2) (1979).

33. See e.g., U.S. NRC, *Solidification of High-Level Radioactive Wastes,* NUREG/CR-0895 (1979); *Hearings on Nuclear Waste Disposal Before the Subcomm. on Science, Technology and Space of the Sen. Commerce Comm.,* 95th Cong., 2d Sess. at 172 et seq. (1978) (statement of Dr. Rustum Roy).

34. Executive Office of the President, *Energy Policy and Planning, The National Energy Plan* (April 29, 1977), p. 70.

35. U.S. NRC, *Regulation of Federal Radioactive Waste Activities,* NUREG-0527, App. F (1979)—hereafter, NRC, *Regulation of Federal Waste;* See also, Helene Linker, Roger Beers, & Jerry Lash, "Radioactive Waste: Loops in the Regulatory System," 56 *Denv. L.J.* 1, 27-28 (1979). For a good description of "spent fuel," see *Minnesota* v. *U.S. Nuclear Reg. Comm'n,* 602 F.2d 412, 413 (D.C. Cir. 1979).

36. See reference 35, p. G-5.

37. 33 U.S.C. Section 1402(j) (1976).

38. 10 C.F.R. Section 50, App. F(2).

39. 10 C.F.R. Section 50, App. F(3).

40. "No one disputes that solutions to the commercial waste dilemma are not currently available. The critical issue is the likelihood (or probability) that solutions, either ultimate or interim, will be reached in time." *Minnesota* vs. *U.S. Nuclear Reg. Comm'n,* 602 F.2d 412, 416 (D.C. Cir. 1979) (Leventhal, J.).

41. See reference 35, pp. 2-1 to 2-2.

42. IRG, *Report,* pp. 9-10.

43. 39 *Fed. Reg.* 32,921 (1974).

44. M. Willrich et al., *Radioactive Waste: Management and Regulation* 62 (1977).

45. NRC, *Regulation of Federal Waste,* p. 2-2.

46. Idem., p. 3-3.

47. U.S. Department of Energy, *Report of Task Force for Review of Nuclear Waste Management* (1978), p. 40.

134   **The Politics of Nuclear Waste**

49. Comptroller General of the United States, *Nuclear Energy's Dilemma: Disposing of Radioactive Waste Safely* (1977), p. 5.

50. NRC involvement in the area of LLW management may increase substantially within the next several years. Proposed regulations are scheduled to be released (adding a new Section 61 to 10 C.F.R. in 1981 or 1982). See U.S. NRC, *NRC Task Force Report on Review of the Federal/State Program for Regulation of Commercial Low-Level Radioactive Waste Burial Grounds,* NUREG-0217 (1977), p. 13; IRG, *Report,* pp. 105-109.

51. 10 C.F.R. Section 50, App. F (1979).

52. Commercial low-level waste burial grounds licensed by states pursuant to the Agreement States program are located at Beatty, Nevada; Hanford, Washington; Barnwell, South Carolina; Maxey Flats, Kentucky; and West Valley, New York. A sixth site, located on state-owned land in Sheffield, Illinois, is licensed by the NRC. See generally U.S. NRC, *NRC Task Force Report on Review of the Federal/State Program for Regulation of Commercial Low-Level Radioactive Waste Burial Grounds,* NUREG-0217 (1977), p. 2. Only the Nevada, Washington, and South Carolina sites are receiving low-level wastes.

53. 10 C.F.R. Section 20.302(b) (1979).

54. 17 U.S. (4 Wheat.) 316 (1819).

55. Idem, pp. 429-430.

56. 254 U.S. 51 (1920).

57. Idem, p. 57. See also *Sperry* v. *Florida,* 373 U.S. 379, 385 (1963); *Leslie Miller, Inc.* v. *Arkansas,* 352 U.S. 187 (1956); *Public Utilities Comm'n* v. *United States,* 355 U.S. 534 (1958).

58. In *Arizona* v. *California,* 283 U.S. 423 (1931), Arizona sought to enjoin the Secretary of Interior from carrying out the construction of Boulder Dam as authorized by act of Congress. Arizona claimed that the Secretary was obligated to submit the plans and specifications for the dam to a state engineer for approval. Rejecting the state's contention, Justice Brandeis, speaking for the Court, held: "The United States may perform its functions without conforming to the police regulations of a state." *Idem.,* p. 451. See also *Hancock* v. *Train,* 426 U.S. 167, 179 (1976) (federal installations not governed by state pollution permit requirements); *EPA* v. *State Water Resources Control Board,* 426 U.S. 200, 211 (1976). See generally, David E. Engdahl, "State and Federal Power Over Federal Property, 18 *Ariz. L. Rev.* 283 (1976); and see Patricia Lucas, "Nuclear Waste Management: A Challenge to Federalism," 7 *Ecology L.Q.* 917, 937-938 (1979).

59. 42 U.S.C. Section 2012 (1976).

60. 10 C.F.R. Section 30, App. F(3).

61. Such an assertion of state regulatory authority is still subject to challenge on preemption grounds. See Section III.C, *infra.*

62. *United States* v. *Georgia Public Service Comm'n,* 371 U.S. 285, 293 (1963).

63. At the present time, all DOE LLW disposal sites are located on federal property and are not subject to licensing either by the NRC or the states. U.S. DOE, *Report of Task Force for Review of Nuclear Waste Management* (1978), p. 40.

64. U.S. Const. Art. I, Section 8, cls. 3 and 11-14.

65. U.S. Const. Amend. X.

66. U.S. Const. Art. VI, C1. 2.

67. Lawrence H. Tribe, *American Constitutional Law* Foundation Press, Mineda, N.Y. (1978) p. 377.

68. For example, *Jones* v. *Rath Packing Co.,* 430 U.S. 519, 525 (1977); *Rice* v. *Santa Fe Elevator Corp.,* 331 U.S. 218, 230 (1947).

69. For example, *Railway Employees' Dept.* v. *Hanson,* 351 U.S. 225 (1956); *Schwabacker* v. *United States,* 334 U.S. 182 (1948).

70. *Florida Lime & Avocado Growers, Inc.* v. *Paul,* 373 U.S. 132, 142-43 (1963); *McDermott* v. *Wisconsin,* 228 U.S. 115 (1913).

71. *DeCanas* v. *Bica,* 424 U.S. 351, 363 (1976); *Hines* v. *Davidowitz,* 312 U.S. 52, 67 (1941); see *Ray* v. *Atlantic Richfield Co.,* 435 U.S. 151, 157 (1978).

72. *Burbank* v. *Lockheed Air Terminal, Inc.,* 411 U.S. 624 (1973).
73. *Amalgamated Association of St., E. Rwy & Motor Coach Emp.* v. *Lockridge,* 403 U.S. 274, 296 (1971); *Pennsylvania* v. *Nelson,* 350 U.S. 497, 504-505 (1956).
74. *Rice* v. *Santa Fe Elevator Corp.,* 331 U.S. 218, 230 (1947); *Southern R. Co.* v. *Railroad Comm'n of Indiana,* 236 U.S. 439, 446 (1915).
75. See discussion at Section I-A, *supra.*
76. 42 U.S.C. Section 2021(c)(4) (1976).
77. 10 C.F.R. Section 50, App. F (1979).
78. See Lucas, "Nuclear Waste Management: A Challenge to Federalism," 7 *Ecology L.Q.* 917, 942-944 (1979); see also Murphy & La Pierre, "Nuclear Moratorium Legislation in the States and the Supremacy Clause: A Case of Express Preemption," 76 *Colum. L. Rev.* 392, 445-46 (1976).
79. S. Rep. No. 86-870, 86th Cong., 1st Sess. 9 (1959); see generally U.S. NRC, Office of State Programs, *Final Task Force Report on the Agreement States Program,* NUREG-0388, Appendix A (1977); Murphy & La Pierre, "Nuclear Moratorium Legislation," note 71a.
80. S. Rep. No. 86-870, 86th Cong., 1st Sess. 9 (1959); see also "Remarks of Mr. Anderson," 105 *Cong. Rec.* 17506-17507 (1959).
81. *Hearings Before the Joint Comm. on Atomic Energy on Federal State Relationships in the Atomic Energy Field,* 86th Cong., 1st Sess. 488 (1959) ("Hearings").
82. *Idem.,* p. 500.
83. S. Rep. No. 86-870, 86th Cong., 1st Sess. 3 (1959).
84. For example, *Northern States Power Co.* v. *Minnesota,* 447 F.2d 1143 (8th Cir. 1971), *aff'd mem.,* 405 U.S. 1035 (1972); *Pacific Legal Found.* v. *State-Energy Resource Conserv. & Develop. Comm.,* 472 F. Supp. 191 (S.D. Cal. 1979), *appeal pending; United States* v. *City of New York,* 463 F. Supp. 604 (S.D.N.Y. 1978); *Marshall* v. *Consumers Power Co.,* 237 N.W.2d 266 (Mich. App. 1975); *Pub. Int. Res. Group* v. *State Dept. of E.P.,* 152 N.J. Super. 191, 377 A.2d 915 (1977); *Missouri ex rel. Utility Consumers Council* v. *Public Service Comm'n,* 562 S.W.2d 688 (Mo. App. 1978).
85. 447 F.2d 1143 (8th Cir. 1971), *aff'd mem.,* 405 U.S. 1035 (1972).
86. *Idem,* p. 1147.
87. Idem, p. 1149.
88. Clean Air Act Amendments of 1977, Pub. L. No. 95-95, 91 Stat. 685.
89. 426 U.S. 1 (1976).
90. 33 U.S.C. Sections 2011 *et seq.* (1976).
91. 426 U.S. at 416-417.
92. See note 77 *supra.*
93. 472 F. Supp. 191 (S.D. Cal. 1979). The federal appeals court, however, reversed the district court in 1981 and upheld the California statute.
94. Cal. Public Resources Code Section 25513.2.
95. 472 F. Supp. at 197.
96. Idem, p. 199-200.
97. Idem, p. 200. As of this writing, a second suit has been instituted in the Eastern District of California in which the validity of Cal. Pub. Res. Code Section 25524.2 has also been challenged. *Pacific Gas & Electric Co. et al.* v. *Cal. Energy Comm'n,* Civ. No. S-78-527-MLR (E.D. Cal.).
98. 463 F. Supp. 604 (S.D.N.Y. 1978).
99. *Idem,* p. 613.
100. 42 U.S.C. Section 2021(k) (1976).
101. S. Rep. No. 870, 86th Cong., 1st Sess. 12 (1959). See also *Northern States Power Co.* v. *Minnesota,* supra, 447 F.2d at 1150.
102. *Marshall* v. *Consumers Power Co.,* 65 Mich. App. 237, 237 N.W.2d 266 (1975). Of interest in this connection is the decision of the California Supreme Court in *No. California Assoc. to Preserve Bodega Head & Harbor, Inc.* v. *Pub. Util. Comm'n,* 61 Cal.2d 126, 37 Cal. Rptr. 432, 390 P.2d 200 (1964), where the court held that the question of the safety of the proposed location of an

atomic reactor to be located at or near an active earthquake fault zone was not a radiological matter which the state was precluded from regulating. See *State ex rel. Utility Consumers Council* v. *Pub. Serv. Comm'n,* 562 S.W.2d 688 (Mo. App. 1978) ("The federal government regulates how nuclear power plants will be constructed and maintained; the State of Missouri regulates whether they will be constructed.").

103. See T. Evans, *Nuclear Litigation,* Thomas W. Evans, Chairman, Litigation Course Handbook Series No. 144, New York Practising Law Institute, (1979) pp. 265-266; *and see* U.S. NRC, *Means for Improving State Participation in the Siting, Licensing and Development of Federal Nuclear Facilities* NUREG-0539 (1979), pp. 1-2. ("States have legitimate interests in repository siting, licensing, and development in the areas of public health and safety, land use planning, economic development, socioeconomic impacts, and transportation routing.").

104. See *No. California Assoc. to Preserve Bodega Head & Harbor* v. *Pub. Util. Comm'n* in note 102 above.

105. *Jones* v. *Rath Packing Co.,* 430 U.S. 519, 525 (1977); *Rice* v. *Santa Fe Elevator Corp.,* 331 U.S. 218, 230 (1947); see *Huron Portland Cement Co.* v. *City of Detroit,* 362 U.S. 440 (1960).

106. See generally, U.S. NRC, *Means for Improving State Participation.*

107. Federal installations have been held to be exempt from local zoning ordinances. For example, *Town of Groton* v. *Laird,* 353 F. Supp. 344, 350 (D. Conn. 1972); see *United States* vs. *City of Chester,* 144 F.2d 415 (3d Cir. 1944).

108. See 42 U.S.C. Section 2012(a), (g)(i) (1976). Citing the Energy Reorganization Act of 1974 and the Department of Energy Reorganization Act of 1977, Professor Lawrence Tribe has recently argued that since the enactment of the Atomic Energy Act of 1954, Congress has radically altered its policy favoring the peaceful applications of nuclear energy. While it is time that Congress has expressed a strong interest in stimulating the developing of so-called alternative energy technology such as solar and geothermal [42 U.S.C.A. Section 7112(6) (Supp. 1978); 42 U.S.C.A. Section 5801(e) (1977)], nothing in the express language of the 1974 or 1977 statutes or in the legislative history indicates that Congress intended to abandon the nation's nuclear power program or phase it out of existence. Rather, to the extent that it can be identified, the congressional policy appears to favor the development of a wide variety of energy sources (including nuclear power) to meet the domestic needs for the remainder of this century.

"While the Federal Government may have effectively abdicated leadership by default in not enacting further legislation clarifying the role of the states in nuclear power development, it has certainly not repeated the mandate of the Atomic Energy Act." Gerald Charnoff & John W. O'Neill, Jr., *Federal-State Regulation of Nuclear Power* in T. Evans, *Nuclear Litigation* (1979), p. 271.

109. *Hines* v. *Davidowitz,* 312 U.S. 52, 67 (1941); see also *Ray* v. *Atlantic Richfield Co.,* 435 U.S. 151, 158 (1978).

110. See L. Tribe, *American Constitutional Law,* Section 6-24, p. 378; *Nash* v. *Florida Industrial Comm'n,* 389 U.S. 235, 239 (1967).

111. See, e.g., 42 U.S.C. Section 2077(a) (1976) (special nuclear materials). "Unless authorized by a general or specific license issued by the Commission . . . no person may transfer or receive in interstate commerce, transfer, deliver . . . any special nuclear material." The ensuing discussion focuses primarily on motor transport. Questions of preemption with respect to other instrumentalities of commerce are not discussed.

112. 10 C.F.R. Section 71.7 (1979).

113. "Memorandum of Understanding - Transportation of Radioactive Materials," 38 *Fed. Reg.* 8466 (1973).

114. *Idem;* see also 10 C.F.R. Section 71 (1979).

115. IRG, *Report,* p. 115.

116. Idem; see *South Carolina Highway Dep't* v. *Barnwell Bros., Inc.,* 303 U.S. 177 (1938); and see *Raymond Motor Transport, Inc.* v. *Rice,* 434 U.S. 429 (1977).

117. 49 U.S.C. Sections 1801 et seq. (1976).

118. 49 U.S.C. Section 1804(a) (1976).

119. *Idem.*

120. 49 C.F.R. Section 178 (1978).

121. 49 U.S.C. Section 1811(b) (1976).

122. *Idem.*

123. *Idem.*

124. Indeed, 49 U.S.C. Section 1811(a) explicitly states: "Except as provided in subsection (b) of this section, any requirement of a State or political subdivision thereof, which is inconsistent with any requirement set forth in this chapter, or in a regulation issued under this chapter is preempted."

125. 49 C.F.R. Section 107.209(f) (1978).

126. Louisiana (LA. REV. STAT. ANN. Section 1072); Minnesota (MINN. STAT. ANN. Section 1161-73); and Montana [MONT. STAT. ANN. Section 75-3-302(1)]. See also 2 Nucl. Reg. Rep. [CCH Section 20.031 (discussing Connecticut's permit restrictions)].

127. LA. REV. STAT. ANN. Section 1072.

128. See Section II.C, *supra.*

129. See *DOT Inconsistency Ruling* (IR-1) (April 4, 1978) (hinting that New York City ban on transport of certain radioactive materials may violate commerce clause).

130. *Philadelphia* v. *New Jersey,* 437 U.S. 617, 623 (1978); *Raymond Motor Transp., Inc.* v. *Rice,* 434 U.S. 429, 440 (1977).

131. 397 U.S. 137, 142 (1970); see also *Hunt* v. *Washington Apple Advertising Comm'n,* 432 U.S. 333, 352-354 (1977).

132. *Philadelphia* v. *New Jersey,* 437 U.S. 617, 624 (1978), citing *H.P. Hood & Sons, Inc.* v. *DuMond,* 336 U.S. 525, 537-38 (1949); *Toomer* v. *Witsell,* 334 U.S. 385, 403-406 (1948).

133. 437 U.S. 617 (1978).

134. Waste Control Act, N.J. STAT. ANN. Section 13:11-1 (West Supp. 1978).

135. 437 U.S., p. 628.

136. 209 U.S. 251 (1908); see also *Reid.* v. *Colorado,* 187 U.S. 137 (1902).

137. 437 U.S., p. 629.

138. But see Lucas, "Nuclear Waste Management: A Challenge to Federalism," 7 *Ecology L. Q.* 917, 940 n.143 (1979).

139. See *Hunt* v. *Washington Apple Advertising Comm'n,* 432 U.S. 333 (1977); *Great A & P Tea Co.* v. *Cottrell,* 424 U.S. 366, 371-72 (1976).

140. Lyons, Kansas (1970) and Alpena County, Michigan (1975-76); see generally Daniel S. Metlay, "History and Interpretation of Radioactive Waste Management in the United States," in U.S. NRC *Essays on Issues Relevant to the Regulation of Radioactive Waste Management,* NUREG-0412 (1978); John Abbotts, "Radioactive Wastes: A Technical Solution?," 35 *Bulletin of Atomic Scientists* 12 (October 1979).

141. Acceding to a request of Senator Pete Domenici, development of DOE's proposed waste isolation pilot plant near Carlsbad, New Mexico was at one time subject to an absolute veto by the state. *Nucleonics Week,* October 12, 1978, p. 3.

142. President Carter announced a national policy on nuclear waste disposal on February 12, 1980. The executive order granted state governments an undefined right of "consultation and concurrence" on decisions respecting the siting of nuclear waste repositories.

5

Consultation and Concurrence: Process or Substance

Emilio E. Varanini, III

This chapter examines the current attempt to develop a process which permits a shared responsibility between federal and state officials for devising methods for the long-term disposal of nuclear waste.[1] The development and implementation of methods for nuclear waste disposal are complicated by the long life of the hazardous waste material and the necessity of protecting the health and safety of current and future generations.

The cardinal proposal of consultation and concurrence is described and then critiqued from the perspective of preexisting state laws and the public position voiced by various institutions, states and multistate bodies. Two case studies are then examined. The first case illustrates the concerns of a potential host state. The second represents the views of an interested state with concerns regarding the existence and demonstration of a nuclear waste disposal technology. The validity of current waste disposal methods are questioned and an alternative to validate nuclear waste disposal is proposed.

Finally, this chapter concludes that before a statutory division of authority between federal and state governments can be determined, many of the generic problems of scientific verification of nuclear waste disposal methods must either be resolved or a process for resolution established by statute.

TECHNICAL AND POLICY CONCERNS

Nuclear waste is a by-product of the fission process utilized in electric power production and in defense plant reactors to produce nuclear weapons material. The waste has been generated since the first reactors went critical in 1943. The basic purpose of a waste disposal technology is to confine and isolate emplaced wastes from the biosphere for very long periods of time.[2] Complete containment is not necessarily required, but elimination of contact with the biosphere or geologic isolation is necessary. There is dispute as to how long this confinement must be maintained. Periods of from 600 to 1,000,000 years have been proposed.[3] The definitive standard for waste disposal technology was set by the National Academy of Sciences: "Unlike the disposal of any other type of waste, the hazard related to radioactive wastes is so great that no element of doubt should be allowed to exist regarding safety.[4]

Although geologic disposal (emplacement at 800 meters or more below the surface) has an intuitive appeal, there are many factors that could contravene the ability of a geologic formation to perform the isolation function. Both the action of emplacing the waste and the presence of the waste may disrupt the very features which give geologic waste disposal its initial intuitive appeal. The emplacement process includes compromising the complex natural pathways from the repository to the surface by investigative procedures, including bore holes and shafts as well as the mining of the repository itself. It has been pointed out that a repository is simply neither the analog nor the obverse of a mine. Natural processes such as seismic events and physical effects such as thermal and radioactive waste rock interaction may affect repository integrity. Future actions by man in search of valuable minerals associated with the repository geology or under the drive of pure scientific speculation may create compatibility or locational concerns about the repository site.[5]

Intuition aside, the risk associated with the development, operation, and long-term isolation function of a repository is dependent on the unique factors of involuntary and intergenerational risk and on the assumptions provided to risk assessment mathematical models. Burckholder and others have made several such calculations, but conclude that the dose to man is dependent on what parameters and assumptions are chosen. In some cases where path length to the biosphere is very short, the largest dose may be up to several thousand time the background dose from naturally occurring radionuclides.[6] In a 1977 article, de Marsily et al. estimated radionuclide releases to the biosphere which in some cases resulted in concentrations several hundredfold higher than the maximum permissible concentrations. These generic calculations show that the safety of disposal is not an inherent feature of geologic disposal, but is circumstantial.[7]

The complexities of developing an acceptable waste management and disposal technology as well as the potential number of jurisdictions affected are illustrated by a simple mathematical construct. If we in the United States establish a nuclear capability of 507 gigawatts capacity by the year 2000, as the government projected a few years ago, a *new* repository would have to open every two to three years after the turn of the century for as long as nuclear power plants remained operational at that level.[8]

There have been at least five specific attempts to solve the disposal problem. All five were abandoned before completion, generally because of unanticipated difficulties. The attempts included: injection of liquid waste into porous media; disposal of liquid waste in solution mine cavity; disposal in vaults in bedrock formations beneath the Savannah River Plant; a repository to accept military waste in Lyons, Kansas; and relocation of a tentative site in a bedded salt formation in New Mexico when that site was determined to contain an unexpected geological problem.[9]

The most relevant example for purposes of understanding current state concerns is that of the Lyons, Kansas experiment. In 1969, plutonium contaminated debris from a fire at the Atomic Energy Commission (AEC) Rocky Flats, Colorado facility was shipped to the Idaho Reactor Testing Station for burial. After strong protests by Idaho legislators, the AEC decided to transform a mine in Lyons, Kansas used in Project Salt Vault (an experimental facility used to test emplacement technology and some near-field effects) into a waste repository and to deposit the contaminated debris. The AEC did so despite the concerns voiced by Dr. William Hambleton, director of the Kansas Geological Survey, that the work done during Project Salt Vault and in other studies could not guarantee safe containment of the wastes in the salt. The events leading up to the AEC's decision are instructive.

> In the light of the survey's concerns, Director Hambleton was invited to serve as a member of the Panel on Disposal in Salt of the National Academy of Science's Committee on Radioactive Waste Management, which had been appointed at the request of the AEC to advise it on waste disposal problems. At a meeting at Oak Ridge, Tennessee, in May 1970 the panel was briefed on the design specifications of the repository. . . . The Kansas Geological Survey was asked to recommend specific studies that would show the feasibility and safety of sinking large amounts of high-level radioactive waste in the underground salt beds. The survey readily complied and submitted its recommendations on June 8, 1970. In the following week, at the request of the NAS Panel, the survey arranged a meeting at the University of Kansas . . . to review the problems of salt storage. While these decisions were continuing, the AEC called a press conference on June 17 and announced the selection of the site at Lyons for a demonstration of waste storage in salt, leading to a permanent development of a national radioactive waste repository."[10]

The AEC decision to establish a repository at Lyons while scientists at the University of Kansas were still planning studies to indicate whether the project was safe brought about a strong protest from the press, the public, and public officials. The AEC attempted to convince these groups that repository design, mining processes, and recovery techniques could be safely engineered. After approximately two years of federal-state conflict, the project was abandoned because the site was determined to be geologically unsafe as a result of salt mining activities which came to light very late in the process. Consequently, after fifteen years and $100 million worth of studies and experiments, the AEC had committed a potentially serious error in judgment because it had not adequately considered local concerns that were as basic as the geologic safety of the site. Only strong state and local opposition to the site brought about serious consideration of the geologic problems.[11]

A PROMOTIONAL/ANALYTICAL CONFLICT

The persistence of the Kansas State Geologic Survey and its superior knowledge of site-specific geology could, of course, be considered a historical event easily overcome in subsequent investigations by the development within the federal establishment of comparative or superior technical capabilities with adequate resources to conduct both generic and site specific geological investigations.

The Lyons, Kansas experience however, may represent a much more universal principle in terms of governance and, therefore, requires analysis by those proposing federal legislation and the concommitant allocation of authority and responsibility for the development and licensure of a geologic nuclear waste repository. The nature of the technical concerns and the level of professionalism in the federal planning at Lyons appears to establish an internal conflict of interest in the relevant federal bureaucracy. It can be argued that the need to resolve doubts or at least show momentum in the waste program simply overcame a more commonsensical and measured approach. As we know now from the Interagency Review Group (IRG) Report, the earth sciences have only been in play within these nuclear waste management programs for a very short period of time.[12]

The explanation for the federal persistence at Lyons might be better explained not by scientific ignorance but by reviewing the policy mission of the federal bureaucracy established to deal with the problem. It is important to recall that the Atomic Energy Commission retained responsibility for research, development, *promotion,* regulation and licensure of waste facilities and generating facilities as well. Not only was there an inherent conflict between licensing and promoting, but an even more fundamental conflict between research and development and promotion once a technology or site was chosen. Thus, whatever waste technology was selected, it would be not only analyzed, but promoted both for its inherent benefit and because it would reduce a major impediment with the potential to slow down deployment or acceptance of commercial nuclear power. Current experience tends to confirm that this conflict continues and to a large extent is one of the more persistent reasons for potential host and adjacent state anxieties.

A recent manifestation of this conflict is found in the Department of Energy (DOE) draft report on the generic environmental impacts in the management of commercially generated radioactive waste. In attempting to establish the basis for comparative analysis among potential technologies, the following data deficiencies were acknowledged:

- "Many of the alternatives as conceptualized in this statement are not in the form of integrated waste management systems and consequently are inherently unsuitable for comparative analysis. It was necessary, therefore, to define the concepts as integrated waste management systems of comparable scope and capacity."[13]

- "Comparative assessment of all possible commercial waste management alternatives was clearly not feasible because of data limitations."[14]

- Insufficient data are available to completely assess several options according to Appendix S methods.[15]

- Only qualitative estimates of two of the critera are made; therefore those critera were not scaled.[16]

- Insufficient information was available for the assessment of two of the criteria.[17] Both of these, critical resource consumption and ecosystem impact, are explicitly required by NEPA.

- With respect to the criterion short-term radiological safety, for which ratings are given, "Only fragmentary information on the number of radiological health effects anticipated from normal system operation was available and virtually no estimates were available on health effects resulting from accident conditions."[18]

- The hazard index approach, proposed in Appendix S, could not be used to assess the long-term radiological safety for any of the options.

- For cost of construction and operation, "estimated costs which are currently available are preliminary and of questionable consistency."[19]

These uncertainties are fundamental because they provide the basis from which DOE exercised its judgment to proceed with the geologic waste disposal method. However, from a technical and legal perspective, the key decisional matrix in the report is flawed because of the admission that values were derived without the benefit of substantive knowledge of the nature on impacts of the alternatives. One possible conclusion is that the report attempts to compare unknowns and that geologic disposal takes on a paramount status because it simply is the only option which has received any appreciable scientific investigation.

If the purpose of this generic environmental impact statement and its conclusions are harmonized, the result is not the DOE objective of pursuing exclusively the geologic disposal option, but instead the selection of an appropriate programmatic strategy. Such a strategy, if fairly taking into account the nature of the unknowns set forth above, would be continuing to pursue safe

short-term storage while gathering the information needed to properly select a
lo::g-term method of disposal.

The policy conflict between the role of analyst, with the complexities set
forth above, and that of promoter comes clearly to light in the summary of the
report, which is digest for generalists, interested laity, and presumably legislative and executive branch decision makers. The summary condenses the discussion to the assertion that "the data indicate that no major obstacles exist to the
successful development of this option (deep geologic disposal) in a safe, cost-
effective and timely manner.[20] A close parsing of this statement appears to
make a virtue out of the absence of information. The numerous concerns and
caveats of the IRG report concerning nuclear waste management are simply
not inducted into this report in any meaningful way. Because of these fundamental concerns about institutional and role bias in the federal government's
administration of nuclear waste management policy and technology, the states
have chosen a diverse manner of counter-strategies and policies to ensure some
degree of leverage in both planning and regulation.

STATE POLICY MOVEMENTS—MAXIMIZED LEVERAGE

Several "models" of state actions necessary to either develop a waste management policy or provide for state and federal interaction in the nuclear waste
management area were reviewed in an earlier paper.[21]

The states can generally be categorized as "interested" or as potential host
states. Interested states are those which have a particular interest in the nuclear
waste management issue in a generic or scientific sense. These states have either
adopted or are considering the California approach in which to some greater or
lesser extent further deployment of nuclear reactors is "coupled" to a state
determination that the federal government has demonstrated a technology for
the disposal of nuclear waste. Some states have implemented this policy
administratively by public service commission decisions that the waste issue
raises the level of economic uncertainty to unacceptable levels and therefore
additional reactors have been ordered withdrawn from utility resource plans.[22]

Potential host states are those where literature searches or preliminary
investigations reveal that the geologic and physiographic features may be
appropriate for a waste disposal site. These states have chosen a greater
number of approaches which include: prohibiting disposal of nuclear waste
within the state; prohibiting disposal of nuclear waste without legislative
approval; establishing administrative agency jurisdiction over nuclear waste
siting; prohibiting importation of nuclear wastes from outside the state;
withholding state-land-use permission for either private or federal nuclear
waste repositories; restricting spent-fuel storage within the state.[23]

Although these tactical considerations may be tested eventually in the courts, they do raise the spectre of inducing into the development of nuclear waste disposal technology a counter-bias to that which may exist or may be perceived to exist at the national level. Thus some of these positions may appear to further cloud technical policy decisions with purely parochial considerations. However, until the processes are exercised, it may be difficult to anticipate specific outcomes.

Even within the categories of proposed or operational state action, the attitudes of the states can be qualified. When analyzing the state views, there are at least four major postures:

- *States which maximize leverage* to participate by both conferring parallel authority to approve or disapprove and which take jurisdiction quite early in the indirect as well as direct geologic reconnaissance phase.

- *States which have designed participatory processes* to coordinate with the federal geologic waste investigations and regulations.

- States which have tied or related resolution of the nuclear waste issues to other nuclear policy questions—the so-called *interested states.*

- *States which are simply captured.* These are jurisdictions where either because of prior commitment to nuclear facilities or because the federal government owns or controls large amounts of land, or both, that state action is relegated to a complimentary and secondary role to that of the federal government.

A PROPOSAL FOR A FEDERAL/STATE MEDIATION PROCESS

Within this milieu, the National Governor's Association has stressed the need to deal with this problem notwithstanding preferences about the future of nuclear power technology because of the present and anticipated waste inventory from commercial reactors as well as past and future waste streams from the defense program. The association places great stress on the interrealtionship of technical and sociopolitical questions, and asserts that

the formulation and implementation of a nuclear waste management program in a timely fashion, to accommodate and ensure public acceptance and sound technical solutions demands the establishment of a participatory process involving federal, state, and local officials, the general public and the private sector.[24]

The National Governor's Association perspective has been fixed largely on process, but it has asserted an underlying governance principle that state and local officials have a responsibility to protect public health and safety in nuclear waste issues.

In August 1978, the National Governor's Association adopted a policy of discharging this responsibility by establishing as a general principle a cooperative federal-state process of approving nuclear waste facilities. However, this general principle was augmented by a process which required state concurrence prior to a final site determination by the Department of Energy.[25] Although this concept contributed heavily to the federal Interagency Review Group on Nuclear Waste Management and was featured in its report to the president, the nature of the technical debate, its detailed determinants, and specific state actions and responsibilities were left to be filled in by subsequent policy developments or in the Congress.[26]

It is important to recall, however, that this desire to find a middle course of shared responsibilities—recoined as "cooperative Federalism"—by both the National Governor's Association and the Interagency Review Group is a new proposal set in a preexisting legal and policy framework. The IRG Report, for example, discusses the current situation with the judgment that "the IRG does not believe that a policy preference for either exclusive Federal supremacy or state veto is appropriate at this time. The IRG does believe, however, and recommends that the consultation and concurrence approach should be adopted."[27]

Regardless of the preferences of the IRG or its desire to avoid a potential stalemate situation, it is pointed out in Professor Harold Green's declaration in the current utility challenge to the constitutionality of the California Nuclear Safeguards Laws that:

It is indisputable that the Federal government has pervasive responsibility and authority with respect to the development of technology for nuclear waste disposal, as well as for the regulation of the handling of such waste and the sites at which such technology is practiced. . . . On the other hand, there are important governmental functions relating to high-level radioactive waste that are beyond the scope of this pervasive Federal regulatory scheme.[28]

In addition, a joint comment by nine states on the draft IRG report to the president asserted the right of a state to reject the siting of a nuclear waste facility within its borders, to enact transportation requirements and stricter routing requirements than those which might be enacted by the federal government, and to recognize the power of the states to link future dependency on nuclear power within the state with the development of a demonstrable method of nuclear waste management. It continued with proposals to develop mechanisms (including financial support) to permit states to participate to a greater extent in licensing and rule-making proceedings before the Nuclear Regulatory Commission (NRC), to recognize the power of the states to veto the burial of certain materials within the state at low-level waste sites, and to develop safety and environmental criteria applicable to nuclear waste facilities.[29] It appears then that the posture which may be taken by a national body representing the several states is to a large extent simply inoperative as real policy because of the diversity among state and local governmental entities. Yet such groups provide a source of guidance to both the executive and legislative branches on structural methods to avoid the potential conflict between national and state interests.

WHAT IS THE CONSULTATION AND CONCURRENCE PRINCIPLE?

There are perhaps two distinct processes which provide a working laboratory concerning the acceptability of the concept of consultation and concurrence. Before detailing these processes, it is necessary to point out that the fluidity of the concept has been almost universally criticized by state and local commentators. To a large extent, the principle simply has no bounds and is so inexact that not only are views about federal supremacy or state retained powers not addressed, but the bases for establishing issues, setting forth the nature of the conflict, as well as conflict resolution mechanisms, are yet to be "filled in."

The IRG report contains the following exchange:

> The IRG was strongly criticized for being insufficiently clear and complete in describing what it meant by these terms [i.e., consultation and concurrence]. The respective roles of the states and the Federal government and the limitations on the authority and responsibilities of each were not defined. Such fundamental questions as the extent to which the states and Federal government have and must themselves exercise the responsibility to protect public health and safety and who represents the State and by what mechanism is state authority exercised were not explored. Most particularly, whether the consultation and concurrence concept implies that states can refuse to permit Federal activities and therefore maintain an effective veto was not clearly stated.[30]

The IRG responded to this criticism with the following restatement of its position:

> The IRG agrees that it failed to articulate adequately what is meant by the terms "State veto" and "consultation and concurrence." By State veto was meant the possibility that a State could at one specific moment—by one of several possible mechanisms—approve or disapprove of Federal site investigation activities or a proposal to site a repository or other facility. The veto concept as used did not include an on-going dialogue and cooperative relationship between the Federal and State authorities.
>
> Consultation and concurrence, by contrast, implies an on-going dialogue participation and the development of a cooperative realtionship between states and all relevant Federal agencies during program planning and the site identification and characterization programs on a regional basis using the systems approach, through the identification of specific sites, the joint decision on a facility, any subsequent licensing process and through the entire period of operation and decommissioning. Under this approach the State effectively has a continuing ability to participate in activities at all points throughout the course of the activity and, if it deems appropriate to prevent the continuance of Federal activites. The IRG believes that

such an approach will lead to better protection of the States' interests than would a system of State veto by which is usually meant that a State approves or disapproves of Federal activities at one specific moment, as well as ensure effective State participation in the Federal Government's waste management program. Such an approach will also lead to freer access to areas for the conduct of geologic investigations.[31]

In a parallel process conducted by the Nuclear Regulatory Commission concerning state participation, representatives of thirty states pointed out substantially similar concerns to those expressed above. Some of the most pertinent conclusions were:

There was general consensus that procedures for participation should distinguish between "host States" and "interested States." Interested States want the opportunity to participate in the development of national policy, procedures and regulations, while host States are interested in evaluations of specific sites in addition to these factors.

There appeared to be consensus among State representatives that the host State should have the right to concur or non-concur in a proposed repository site but there was no clear consensus as to how a non-concurrence could be arbitrated or subjected to rejection by another authority (e.g., Congress, the President, an "Executive Planning Council," etc.)[32]

The two laboratories where the concept of consultation and concurrence is being affected are in the legislatures of the several interested and potential host states and in the executive branch of states where the legislatures have acted.

Although consultation and concurrence proposal has been criticized for a lack of specificity in several states, it is being fleshed out on a de facto basis. These states have adopted legislation to guide the executive branch and have active programs underway. Although the state programs are in large part embryonic, it is still possible from official statements and postures driven by federal pacing events, such as the IRG report to the president, the draft Waste Isolation Pilot Plant (WIPP) environmental impact statement (DEIS); and the draft environmental impact statement for the management of commercially generated radioactive waste, to examine early concerns with cooperative ventures between the federal government and the states. In this case, comments of New Mexico, the leading potential host state, and California, an interested state, are instructive.

CAN POTENTIAL HOST STATES INFORMALLY DEVELOP A FEDERAL/STATE ROLE?

The process of consultation in the case of New Mexico apparently spans a broad spectrum of concerns. However, on a generalized basis, it may be summarized by the necessity of completing the data base required to make judgments on areas as diverse as transportation planning, geologic factors potentially affecting public health and safety, and local socioeconomic impacts before proceeding with the project or seeking its approval.

In his letter to Secretary Duncan transmitting the state's review of the draft WIPP EIS, Governor Bruce King pointed out that the final EIS date was not coordinated with state studies both on socioeconomic effects, and on technical and scientific issues and questioned how the consultation and concurrence process would be utilized to incorporate these views. He asserted that if the final EIS itself were not properly supplemented, then he would declare it inadequate. Consultation and concurrence, which the governor stated was still being defined, would then assume not only the state role in determining final project scope (i.e., differentiating among project objectives, including a TRU waste repository, high-level waste research and development projects, and spent-fuel respository), but all health and safety concerns as well. Finally he pointed out that although the EIS is certainly an integral component of consultation and concurrence, even a state sign-off on the EIS would not constitute final concurrence.[33]

The WIPP DEIS comments by New Mexico were a virtual litany of concerns with state officials expressing reservations about transportation, emergency planning, and socioeconomic effects in particular. The concerns, however, when generalized or simply stated above, appear to be part of a quite normative, iterative process between state and local officials. However, on detailed review of the comments, fundamental institutional and technical conflicts are identified.

The first stated concern was the Department of Energy's inability to take a firm position on NRC licensure of the WIPP facility. Because congressional committees intervened in this matter, particularly the House Armed Services Committee, DOE was reluctant to take too visible a position in this controversy which is of grave concern to the state of New Mexico. If licensure does not take place, what proxy for it will be developed? What does that do to the normalized responsibilities of the Environmental Protection Agency's health and safety criteria; the Nuclear Regulatory Commission's standards to meet them; and the Department of Energy's role as an advocate of a given waste technology?

Secondly, what type of standards will be worked out in transporting waste, and, more importantly, which entity will enforce them? How will current laxity of enforcement be improved?

Lastly, what type of liability will be imposed, and is Price-Anderson liability limitation and indemnification in effect? What if that is unacceptable to New Mexico and becomes a matter of controversy?[34] Even these issues pale against the concerns raised by the Governor's Advisory Committee on WIPP. These issues are so fundamental that the salient points are set forth as presented.

Because of the limited scientific information available in the draft DEIS and its references in certain areas, it is not appropriate to imply as is done by omission in Chapter 10, that there will be no long-term adverse impacts on the proposed WIPP respository. . . . A decision to proceed with the construction of WIPP should be supported by stronger scientific arguments than are given herein.[35]

The committee recommended further that the Department of Energy be requested to comply with all appropriate state regulations, a somewhat novel approach to consultation.

Finally, the committee raised several issues about unresolved geologic aspects of the site which, if not properly understood and applied in the decision to proceed, could possibly result in severe repository failure.

The report sets them out as follows:

- What is the origin, evolution, and occurrence of the high-pressured brine reservoirs which were encountered in the upper part of the Castile formation at ERDA No. 6 and in several oil wells in the area?

These high pressure fluids (1,800 psi at ERDA No. 6) have occurred in the unit which is immediately below the repository level proposed for remote handled high-level waste.

- What is the origin and the occurrence of the "breccia pipes" which have been encountered in the area?

These may be localized deep-dissolution features which originate in the lower portion of the evaporils and migrate upwards. Such localized dissolution features could now exist or develop later beneath the site.

- What are the processes and the rates of dissolution of salt near the site?

Deep-dissolution within the evaporite sequence is especially critical because there appears to be preferential removal of the salt horizon which is proposed for the repository.[36] The nature of these concerns are then stated in a series of candid possibilities including in the case of brine reservoirs or developing breccia pipes a potential threat to the *immediate* as well as long-term integrity of the site. The production of gas (2,000 psi) from transuranic waste is cited as a potential cause of fracturing the overlying rock with direct release of contaminants to the atmosphere, and the present uncertainties about ground water hydrology are pointed out as limiting the capability to do risk assessment.[37]

These matters are not set forth here as assertions that such matters will be found to be the predominant scientific view or whether, subject to further analysis, they can be adequately dealt with or even dismissed. The critical point is that consultation and concurrence must be contorted to encompass all of these types of issues, and the probability is that from time to time, nonconcurrence postures may be advanced.

INTERESTED STATES: GENERIC SCIENTIFIC VERIFICATION OF TECHNOLOGY FEASIBILITY

In the case of California, state law couples the status of demonstrating high-level nuclear waste management technology with the further deployment of nuclear reactors. A large number of papers explicating the California position and review of the technology have been prepared.[38] For the purpose of this chapter, the important point is that as an "interested state," California, in both public forums and direct consultation with the Department of Energy, has proposed an alternative, or "California," approach to demonstrating a nuclear waste technology. The details of the consultation are contained in the appendix to this chapter. This state consultation points out in particular the need for a parallel program to demonstrate:

• A fundamental understanding of all phenomena relevant to nuclear waste isolation from the environment

• A series of tests confirming that the postulated characteristics needed to isolate wastes are in fact effective

• Validated predictive models of repository performance, radionuclide transport, and exposure/dose

• Criteria and standards defining the necessary and sufficient conditions required of a repository

• Confirmation that at least one site meeting the criteria and standards exists[39]

A FORMULA TO REDUCE INTERESTED STATE CONCERNS

Because of the variation of interested states' concerns, as well as the nature of the problem of nuclear waste management itself, it is necessary to discuss to some extent the technical and policy determinants which should be considered. Since technology is initiated from scientific to practical and routine applications, it may be desirable to review analogs in existing federal laws and court decisions involved with technology in order to understand the key determinants associated with validating a technology. At least in this sense, the roles of federal and state governments can be designed with some understanding of what must be accomplished. This, of course, presupposes that the concerns of interested states would be absorbed into federal legislation. Legislation in other areas which has dealt with not only technology applicability but technology forcing functions includes the Clean Air Act, Federal Water Pollution Control

Act, the Noise Control Act, the Occupational Safety and Health Act, and the National Traffic and Motor Vehicle Safety Act. Court decisions interpreting the Patent Act as well are instructive, particularly in regard to issuing patent protection upon the operability of the intervention and its reduction to practices.

In general, technologies have been determined to be effective, available or currently available, demonstrated or adequately demonstrated, practicable, economically achievable or economically reasonable, feasible or technically feasible, operable or reduced to practice under following conditions:

1. Successful testing
   a. At full scale, full size, full capacity or large capacity
   b. Under expected operating conditions, including start-up, input variations, output fluctuations, process interruptions, imperfect maintenance, malfunction of related systems and unusual environmental circumstances
   c. Documenting the absence of offsetting disadvantages (environmental, cost, or other)
   d. Over expected operating lifetime, if possible
2. Successful operation, demonstrating
   a. Technical feasibility
   b. Economic praticality
   c. Environmental acceptability
   d. Reliability

The facts and evidence necessary to make these determinations include tests of operational facilities as well as unconditional guarantees.

A suggested set of criteria can be developed from these types of considerations:

1. A technology "exists" if there is a technique or method upon which enough information is available from testing at bench, pilot, or semi-work scale under conditions approximating those under which it must perform in actual use to establish convincingly that the technology can achieve its intended purpose consistently without prohibitive offsetting disadvantages. This definition is similar to patent law concepts of utility, operability, and reduction to practice.
2. A technology exists for the construction and operation of a plant if the technology has been comprehensively tested under expected operating conditions, including start-up, input variations, output fluctuations, process interruptions, imperfect maintenance, malfunction of related systems, and unusual environmental conditions at a reasonable scale (either from actual full-scale operation or a pilot or semi-works scale plant) to establish con-

vincingly that the technology can operate consistently, reliably, routinely, and within prescribed occupational and public safety, emissions and safeguards limits over extended periods of time. Existence of a technology for operation of a plant is closely related to an "available, practicable technology" as used in the Clean Air Act.

3. A technology has been developed and demonstrated that it can achieve its intended purpose reliably, without offsetting disadvantages such as threats to occupational or public safety, environmental quality or national security, and within prescribed safety, emissions and safeguards limits. These qualities must be shown to be achievable through previous operation of the technology or through comprehensive testing at a pilot or prototype scale under conditions approximating actual use including abnormal, upset, or unusual conditions.

CONCLUSION

The issues and concerns in nuclear waste management transcend the status of any state or states. The "interested" state as well as potential host states will require approximately the same level of scientific verification in order to determine that the resolution of the nuclear waste management issue is either being reasonably pursued or is in a condition which would allow siting or operation of a repository to proceed. The "common mode failure" of this process appears to be the dualistic nature of the U.S. Department of Energy which is assigned an analytical or developmental and a promotional role in the process.

The absence of federal legislation concerning the responsibilities of the states has resulted in a series of creative attempts by various states to carve out either a complementary or an independent regulatory role in order to maintain leverage to participate in the decision process. Federal legislation which does not set an objective process to deal with these generic as well as site-specific questions, but which provides a structured ambiguity about roles, powers, and duties as between the states and federal government, will continue the current policy stalemate and may create an even greater climate of mistrust. Specific scientific criteria and clear and concise roles for state and federal governance could shift the DOE role toward analysis and scientific verification and away from promotion while providing a clear set of statutory directives in the event that ultimate supervision is executed by the courts.

APPENDIX
CONSULTATION WITH U.S. DEPARTMENT OF ENERGY: AN ALTERNATIVE CALIFORNIA APPROACH TO DEMONSTRATING HIGH-LEVEL NUCLEAR WASTE TECHNOLOGY

The California approach is not a solution to the waste management problem. However, it can be instrumental in determining whether geologic disposal is a possible solution. Its objective is to show that the process of modeling repository system performance and of predicting the development of the repository with time can be done with sufficient accuracy to accept models, as applied by scientists and technologists, as demonstrations of repository safety. The process of modeling involves selecting a computer code which includes at least the mechanisms of interest; setting up the system to be analyzed as defined by the system boundaries, the geologic structure, and the repository structure; describing the response characteristics of the various materials present in the system, including material properties; determining the initial conditions at the start of an experiment; and predicting the conditions at the end of an experiment.

The modeling process will be successful in representing reality only if the underlying mechanisms responsible for earth science and materials science phenomena are understood, if this understanding can be reduced to mathematical relationships or constitutive equations, and if the initial and final conditions can be measured and specified. If the modeling efforts are successful and if the experimental testing of the models is stringent, then there can be a reasonable degree of assurance that relevant phenomena are understood well enough that credible long-term predictions can be made. The long-term predictions will always contain uncertainty; the California approach is proposed as a means of attempting to reduce that uncertainty to a level consistent with current capabilities.

There are two aspects to establishing the feasibility of the geologic disposal concept. The first is the acquisition of sufficient understanding of earth and materials sciences phenomena in order to make credible predictions of systems response over time under both "normal" and "abnormal" or accident conditions. The second is convincing the public that nuclear waste disposal can be done with safety; that is, achieving a "social consensus" in the terms of the Interagency Review Group on Nuclear Waste Management (IRG).

Although we agree in essence with the IRG on the elements of a conceptual demonstration of the geologic disposal concept, we regard them as being incomplete in that they may not lead to a social consensus unless the proponent agency recognizes that both the substance (technology) and the process (decision) must be accepted. In our opinion, the existence of the following elements would constitute a demonstration of the safety of isolation:

1. A fundamental understanding of all phenomena relevant to nuclear waste isolation from the environment.
   - Hydrology—ensuring the absence of circulating groundwater.
   - Seismic and tectonic activity—avoiding areas where earth movement and shaking

is likely. Geologic stability is one of the commonly proposed site suitability criteria. However, past geologic stability is no guarantee of future stability.
- Waste form chemical and physical properties—selecting solid forms that have desirable properties in a selected waste management system.
- Waste-rock interactions—ensuring that the physical, chemical, and thermal effects induced by the emplacement and the presence of the waste do not create unmanageable disruptions.
- Sorption (retention) phenomena—having surrounding strata as a barrier through ion exchange, filtering, etc., to "hold-up" any waste radionuclides which may be released from the waste form.
- Rock mechanics—ensuring the structural integrity of the formation. The processes of finding, testing and using a rock formation for a repository represent a disruption of a site that was selected because of its desirable features in an undisturbed state. At issue is the geologic and hydrologic environment of this disturbed rock, not that of the virgin rock. Past predictions in engineering geology have not been particularly successful
- Shaft sealing and borehole plugging—being able to reseal all penetrations and restore the original long path from the formation to the biosphere. The path length along a shaft or borehole is straight and short. The technology for reliably sealing boreholes remains to be developed.
2. A series of tests confirming that the postulated characteristics needed to isolate wastes are in fact effective.
3. Validated predictive models of repository performance, radionuclide transport, and exposure/dose.
4. Criteria and standards defining the necessary and sufficient conditions required of a repository.
5. Confirmation that at least one site meeting the criteria and standards exists—a critical task recognized by the National Academy of Sciences in 1957.

We would prefer that the IRG had listed an ongoing R&D program first rather than last. The first, second, and third elements that we list describe the type of R&D program that we think needs to be done. By not emphasizing the central place of the R&D program in achieving a social consensus, the IRG report can give the false impression that the problem of nuclear waste management is institutional rather than technical. However, at the present time, not even the scientific feasibility of the conventional geologic concept has been established. This must be done before the many institutional problems will begin to vanish.

The IRG implies that the public would find the current state of knowledge acceptable if they were only aware of it; whereas we maintain, on the basis of a careful review, that the current state of knowledge is inadequate to satisfy the public and thereby lead to a social consensus.

We agree with the IRG that there must be an R&D program to increase the state-of-the-art of knowledge. At issue, though, is the timing of experiments and the scale of those experiments. The California approach emphasizes vault tests. Vault tests are the most sophisticated and, if properly designed and phased with smaller-scale experiments, provide the most convincing results to observers.

Longer-term tests, so that equilibrium or steady state conditions are achieved, are required on topics such as fluid flow through fractured rock, rock thermal mechanical and compositional response, and the geochemical retardation of radio-nuclides moving in solution. After normal test conditions, some experiments should be run to failure, that is, overloaded, not to simulate possible future conditions, but rather to determine if the observed response can be predicted accurately.

We believe that experiments can be designed not only to check near-field thermal effects, as has been done in the past, but also to determine the effectiveness of postulated geologic barriers and of engineered barriers to the release of radionuclides, to quantify hold-up factors for radionuclide transport, and to confirm the ability to seal shafts and restore the integrity of the geologic formation.

The concern of the public is not so much with the repository environment itself, but rather with the release of radionuclides from that environment. Therefore, it will be necessary to concentrate on phenomena that become manifest in the intermediate field and far field. This has implications for both the type and duration of experiments. One- and two-year long experiments will not be particularly convincing to the public, because the period of concern is thousands of years.

Our generalized vault testing recommendations are as follows:

- Vault tests be run in at least two different media and at a minimum of two sites per each promising medium.

- Tests be run for several years such that (a) equilibrium or steady state conditions are attained, and (b) effects and responses have some opportuinty to manifest themselves.

- Tests be run for several years such that (a) equilibrium or steady state conditions are attained, and (b) effects and responses have some opportuinty to manifest themselves.

- Some thermal experiments be run at well above design basis heat loads to determine if unexpected effects occur and to test our ability to predict thermal response.

- Large-scale failures be induced both structurally and thermally to assess the full range of predictive capabilities of design analyses for long term isolation. One objective would be to provide realistic criteria for the definition of permissible stresses, deformations, temperatures, and fluid flow rates.

- Water be added to the formation to simulate effects of repository flooding.

- Experiments be performed to evaluate the role, response, and effectiveness of engineered barriers.

- Simulated wastes be emplaced as part of a source term characterization program.

- Spent fuel be emplaced to permit the observation of effects of combined thermal and ionizing radiation. A condition is that retrievability capability never be lost.

Our recommendations for vault testing include more than the types of experiments to be performed; they include also recommendations for the scientific environment in which experiments are planned, executed, analyzed, and interpreted.

There should be a clear statement of what an in situ experiment is to accomplish. The experiments we have in mind will be designed to validate models and predictive abilities, not only to measure something, to gather data, or to increase understanding; these later objectives are subsidiary to the major objective. That is, the objective is to test the current level of understanding. The steps in the process of experimental design would proceed something like this:

1. Determine which validation experiments are most properly done in a laboratory and which can be done only in situ.
2. Determine the maturity of understanding with respect to a proposed in situ experiment. Because of their cost, in situ experiments should be performed only if the anticipated output justifies the expense.
3. Determine experimental scale with respect to geometry and time and determine the distance over which observation should occur.
4. Prepare a written statement of experimental design. Maintain firm, centralized managerial control and direction over the course of the experiment.
5. Predict the experimental results expected. Pre-experimental issues such as the acceptable accuracy and precision, the location for and type of experimental observations, and the implications of observations outside of predictions should be resolved. An experiment that gives a result identical to prediction is probably of least interest. Note that we propose that the objective is to test understanding, especially of mechanisms.
6. Maintain an independent peer review over the execution of the experiment and the analysis of the data. One purpose of the peer reviewed process is to serve as an *independent* error-detection and correction mechanism.
7. Assess how short-term experimentation relates to the ability to ensure long-term isolation of radionuclides.
8. Present the results in public forums with honest appraisal of successes and failures.
9. Factor the experience into future experiments. Assess how the experiments might have been done differently to have been more effective.

Our recommendations call for the rigorous application of the scientific approah of hypothesis formulation and validation by means of a proven understanding of relevant natural phenomena. Minimum program requirements include repository performance modeling and validation, in situ experiments rather than laboratory experiments, and independent peer review of research and development program design, execution, and interpretation.

The California approach has the following significant advantages over the current DOE program:

- Provides rock mechanics data useful for repository construction and operation prior to the beginning of repository design work

- Identifies problems early on so that solutions can be sought prior to or concurrent with repository operation

- Quantifies advantages and disadvantages of rock types so that the best medium can be selected for the second repository, if not the first

- Provides knowledge base for establishing defensible site suitability and repository performance criteria and standards

- Facilitates the regulatory process

- Signals a commitment by the federal government to solve the nuclear waste management problem

- Establishes technical and public credibility in the geologic disposal concept

Another feature of the California approach is that it will provide the above advantages well before the DOE program will. The first repository may not be available before 1997, if then. If existing mined cavities can be found and obtained for experimentation, then vault testing could commence within a few years, a decade before a DOE repository-based program could get underway. We commend the DOE for using the existing opportunity presented by the Stripa iron mine in Sweden and encourage the DOE to continue to use Stripa and other mines for large-scale experimentation within a carefully planned program.

One difficulty with the DOE approach of deferring certain experiments until the first repository is available is that certain experiments may be precluded by the desire to maintain isolation. On the one hand, there is a desire to learn as much as possible about a repository site in order to provide the input data for a safety assessment. If the site is to be used also for vault testing, then disruption of the formation beyond that caused by mining will be required for both experimentation and observation. On the other hand, there is a desire to disrupt the site as little as possible lest the site characteristics favoring isolation be destroyed. The manner in which this conflict will be resolved has not been adequately discussed in the past.

The vault test approach meets the regulatory criterion set forth in the IRG subgroup I report (pp. 45-46). That regulatory criterion is not predicated on beliefs, but includes facts and conclusions necessary to discharge fundamental responsibility. The "uncertainties" set forth include (a) whether the mined repository *concept* and any particular site and design will be sufficient to meet the overall EPA safety standards, and (b) the fact that until the NRC regulatory guides are available, DOE cannot be certain about the type of information and analysis to be required by NRC.

The outcome of the licensing process cannot be known for certain in advance. It seems quite clear that if basic knowledge is not at the threshhold of the licensing process, then the process will be compromised and the licensing will be perceived as simply marketing. If this is not the case, the licensing process will identify the lack of scientific support for the concept and licensure will not proceed, resulting in a substantial delay in finding acceptable waste disposal solutions and severely damaging public confidence.

NOTES

1. This paper was written prior to the formation of the State Planning Council on Radioactive Waste Management in February 1980. Although the council has discussed consultation and concurrence and has passed resolutions expressing the council's views on some components of the consultation and concurrence process, a great deal of work remains before the process is accepted nationally and implemented as part of repository site selection. Thus, the events which have occurred between this paper's presentation and its publication do not alter the relevance of its content.

2. Emilio E. Varanini, III, *Comments on the Adequacy of the Draft Statement for the Management of Commercially Generated Radioactive Waste,* submitted to Division of Waste Isolation,

U.S. Department of Energy, San Francisco, October 4, 1979; U.S. Energy Research and Development Administration, *Alternatives for Managing Wastes From Reactors and Post-Fission Operations in the LWR Fuel Cycle*, ERDA-76-43, May 1976; Jet Propulsion Laboratory, *An Analysis of the Technical Status of High Level Radioactive Waste and Spent Fuel Management Systems*, prepared for CERCDC, JPL Publication 77-69, Pasadena, December 1, 1977.

3. Examples of various proposals are given in American Physical Society, *Report to the American Physical Society by the Study Group on Nuclear Fuel Cycles and Waste Management*, July 1977, p. S112. One proposition is that of Willrich et al. (1976): "because the [ingestion] waste hazard index does not fall below that of pitchblende [natural ore with 60 percent uranium] until about one million years, these [LWR] wastes . . . must be contained and isolated for as long as one million years." (APS, p.S112) Another rule of thumb is to allow for 10 half-lives of decay for a given radionuclide, or a reduction in radioactivity by a factor of 1024. However, the question then is: for which radionuclide is the rule to be applied—10 half-lives of Pu-239$_5$ (244,000 years), of I-129 (1.7 10^8 years), or of Th-230 ($8^2 \times 10^5$ years)?

4. U.S. National Academy of Sciences, National Research Council Committee on Geologic Aspects of Radioactive Waste Disposal, *Report to the Division of Reactor Development and Technology*, U.S. Atomic Energy Commission, May 1966, p. 3.

5. Emilio E. Varanini, III, *Status of Nuclear Fuel Reprocessing Spent Fuel Storage and High-Level Waste Disposal* - Draft Report, California Energy Commission Nuclear Fuel Cycle Committee, January 11, 1978.

6. H. C. Burckholder et al., *Incentives for Partioning High-Level Waste*, (Seattle: Battelle Pacific Northwest Laboratories, BNWL-1927, November 1975), pp. 21-38.

7. G. de Marsily, E. Leodus, and J. Margat, "Nuclear Waste Disposal: Can the Geologist Guarantee Isolation?" *Science* 197 (August 5, 1977): 519.

8. Varanini, *Status of Fuel Storage and Waste Disposal*, p. 218.

9. Ibid., pp. 123-127.

10. Richard S. Lewis, *The Nuclear Power Rebellion* (New York: Viking Press, 1972), p. 159.

11. Richard S. Lewis, "Kansas Officials Oppose AEC on Radioactive Waste Repository, *Nuclear Industry* 18. (March 1971); and U.S. Congress, *Joint Committee on Atomic Energy Authorization Legislation: Fiscal 1975, 93rd Congress*, 2nd Session, 1974; and William W. Hambleton, "The Unsolved Problem of Nuclear Wastes," *Technology Review* 74, March/April 1972.

12. U.S., Department of Energy, *Report to the President by the Interagency Review Group on Nuclear Waste Management*, Washington, D.C., March 1979.

13. U.S., Department of Energy, *Generic Environmental Impact Statement: Management of Commercially Generated Radioactive Waste*, April 1979, p. 4.2.

14. Ibid., p. 4.3.

15. Ibid., p. 4.9.

16. Ibid., p. 4.10

17. Ibid.

18. Ibid., p. 4.13

19. Ibid., p. 4.17

20. Ibid., p. 1.23

21. Emilio E. Varanini, III, *A View on Consultation and Concurrence* (Battelle: Human Affairs Center, September, 1979).

22. Ibid, p. 9.

23. Ibid., pp. 10-11.

24. National Governor's Association, *Toward Establishing a Responsive and Acceptable National Nuclear Waste Management Policy*, briefing paper prepared for the Nuclear Waste Management Workshop Participants, April 23-24, 1979.

25. National Governor's Association, Nuclear Power Subcommittee, *Recommendations Toward Establishing A Publicly Responsive and Acceptable National Nuclear Waste Management Policy*, May 8, 1979.

26. U.S., Department of Energy, *Report to the President by the Interagency Review Group on Nuclear Waste Management,* Washington, D.C., March 1979.

27. Ibid., p. 93.

28. *Pacific Gas & Electric Company, et al.* vs. *State Energy Resources Conservation and Development Commission et al.* United States District Court Eastern District of California, No. S-78-0527 MLR, October 29, 1979. See especially, summary of declarations by Harold P. Green, pp. 7-9.

29. California, Energy Resources, Conservation and Development Commission, Nuclear Fuel Cycle Committee, *Comments on the Draft Report to the President by the Interagency Review Group on Nuclear Waste Management,* December 1, 1979.

30. Ibid., p. 94.

31. Ibid., p. 95.

32. U.S., Nuclear Regulatory Commission, *Means of Improving State Participation in the Siting, Licensing and Development of Federal Nuclear Waste Facilities,* A Report to Congress, Washington, D.C., March, 1979.

33. See letter of transmittal from Governor Bruce King in U.S., Department of Energy, *Radiological Health Review of the Draft Environmental Impact Statement: Waste Isolation Pilot Plant,* of Robert H. Neill, et al., Nuclear Assessments Office, Environmental Evaluation Group, Environmental Improvement Division, Health and Environment Department, State of New Mexico, September 6, 1979.

34. Ibid., Transmittal letter from Governor Bruce King to Secretary of Energy, Charles Duncan, p. 3.

35. Ibid., p. 3.

36. Ibid., p. 6.

37. Ibid., pp. 9-11.

38. Examples cited earlier in this paper include: *Comments on the Adequacy of the Draft Statement for the Management of Commercially Generated Radioactive Waste, Comments on the Draft Report to The President by The Interagency Review Group on Nuclear Waste Management.* Also see E. E. Varanini, *Comments on the Adequacy of the Draft Environmental Impact Statement Waste Isolation Pilot Plant,* California Energy Commission, September, 1979.

39. E. E. Varanini, III, *Status of Nuclear Fuel Reprocessing Spent Fuel Storage and High-Level Waste Disposal,* California Energy Commission, January 11, 1978.

6

Public Participation:

U.S. and European Perspectives
Dorothy S. Zinberg*

It is almost one hundred years since William K. Vanderbilt, annoyed by a reporter's inquiry about whether the needs of the public had been considered in the siting of railroad stations, sputtered, "The public be damned." Today nuclear industrialists and government officials engrossed in their own "siting" problems may harbor similar regressive sentiments, but they would be loath to express them aloud. Rather, they and their colleagues in most other democratic countries have for the past decade been insisting that were the public sufficiently informed, it would support government nuclear policies. Their prediction seems optimistic when considered in light of the public's growing fear of radiation, alarm over cases of past governmental irresponsibility, and suspicion that official disclosure is still incomplete.

Civic unease has been aggravated by the publicized assessments of scientists whose disagreements with industry and government provide such a range of views that by selection, any position can be supported. Seeing no consensus among experts, the public has begun to insist not only that it be educated, but also that it be allowed to participate directly in decisions regarding commercial nuclear power. The stage has become crowded with new actors—environmentalists, public interest groups, labor unions, local citizens' groups, ideologues, ethicists, and anarchists—and the nuclear issue has become theater—a serialized drama ready-made for the media.

No small part of the public's heightened awareness of nuclear energy issues has been due to the media portrayal of worldwide nuclear events. Reports from

* Sections of this paper were presented under the title "Nuclear Wastes and Future Generations—Uncertainty Compounded," at the Spring Hill Foundation Meetings in Minnesota, April 1980. The research was supported by the Department of Energy (DOE 506-7058) and by the Office of Technology Assessment (OTA 506-7179).

aware of nuclear tensions in other countries, while the towers of Three Mile Island (TMI) have become a universally recognized symbol. For those who support nuclear energy, TMI epitomizes safety—no one was killed and the locations such as Gorleben, Plogoff, and Dounreay have made Americans release of radioactivity was slight. For those opposed, it has become a ground on which to fight the battle to close down the nuclear establishment.

Three Mile Island (March 1979) returned the issues of nuclear power plant siting and safety to public prominence. For five years preceding the event, radioactive waste management had been identified in both the United States and Europe as the one problem that would have to be solved if nuclear energy were to be ultimately acceptable.[1] Although at first glance public scrutiny seems wholly fixed on safety and siting issues, a more careful reading suggests that waste management issues are being examined as an aspect of safety—not overlooked. In fact, during the past two years, radioactive waste management has taken on two new dimensions. It is now identified with the larger, generic problem of toxic and carcinogenic chemical waste management, which parallels the problems of low-level radioactive waste disposal. The plight of the residents of Love Canal in Buffalo, and reports from West Valley, New York of birth defects have further intensified public anxiety about the effects of improperly stored toxic chemical wastes, whether radioactive or not. In addition, as information about long-term negative effects of mismanaged radioactive materials has begun to accrue, church organizations have taken up nuclear waste issues. They question the ethical justification of leaving radioactive material for future generations to manage.

Even if radioactive waste problems are temporarily merged with concern about the safety of commercial nuclear reactors, their successful resolution will continue to be crucial for public acceptance of nuclear power. Public interest groups and informed citizens are demanding that government and industry provide to the public the information and reassurance it requires before accepting nuclear power.

There is a growing national trend of opposition in the United States to big government, large-scale technology, and the centralization of authority, and there is general skepticism about the government's ability to deal with the problems of society. Yet these same elements must play a central role in the resolution of societal problems. It is urgent that the tension between the federal government and the individual citizen be resolved through the intermediary organizations of state and local governments, through advocacy groups, and through direct participation. For unless workable policies are adopted, the country may face a de facto moratorium on many new technological ventures.

This tension is evident in the continuing controversy surrounding nuclear waste disposal. Nuclear waste, the Achilles heel of the nuclear fuel cycle, symbolizes the vulnerability of the entire fuel cycle to public antipathy. In at least twelve states, public-initiated legislation has imposed restrictions on nuclear waste storage, while in several European countries the courts have decreed that no new nuclear power plants can be built until the management of nuclear waste has been resolved. Nuclear waste management is testing the process of government in a democratic society, from policymaking through management and regulation.[2]

Because the present stalemate is widely believed to stem in large part from years of inadequate planning, poor management, and a lack of sensitivity to the concerns of the public, it is useful to examine the past in order to plan for the future.

PAST U.S. GOVERNMENTAL EFFORTS TOWARD A NUCLEAR WASTE MANAGEMENT POLICY[3]

Nuclear wastes have been accumulating for thirty-five years. Yet only in the past decade has serious attention been given to the development of a waste management policy. To understand this seeming negligence it is necessary to view the problem of nuclear waste and spent-fuel management as having originated with the production of bomb-grade material in 1945. Military wastes caused little concern at first because the producers naturally recognized the overriding urgency of developing weapons for the war. The volume of waste was small, and its disposal appeared to be a minor problem relative to the enormous task of bringing a major war to an end. Furthermore, the rapid success in building nuclear weapons contributed to a psychological state of technological euphoria. Those responsible for harnessing nuclear energy were quite certain that appropriate waste disposal technologies would be developed when needed.

After the war, nuclear scientists continued to focus on the front end of the fuel cycle. Carroll L. Wilson, who served as the first general manager of the Atomic Energy Commission (AEC) from 1947 to 1951 and later as professor of technology at MIT, reminisced about what went wrong in the nuclear industry's planning for waste management:

> Chemists and chemical engineers were not interested in dealing with waste. It was not glamorous; there were no careers; it was messy; nobody got brownie points for caring about nuclear waste. The Atomic Energy Commission neglected the problem. Very little solid work was done on the conversion of high-level waste into solid non-leachable form nor on how to store it. It is true that a program of studies led to selecting a Kansas salt mine as a storage place but it turned out that it was not suitable. The central point is that there was no real interest or profit in dealing with the back end of the fuel cycle."[4]

The impressive wartime accomplishments as well as the safety record of the Manhattan Project seemed to prove that scientific talent, combined with substantial financial support, could solve previously insuperable problems. Considering past achievements, their confidence was understandable.

Not until passage of the Atomic Energy Act in 1954 did the AEC, established to deal with both military and civilian nuclear power, undertake a serious examination of waste disposal. At the AEC's request, the National Academy of Sciences (NAS) began a study of the feasibility of disposal in geologic formations. The NAS committee recognized the potential health hazards, the technical problems, and even the political questions associated with this issue. Their report included this correct prediction: "Disposal is the major problem in the future growth of the atomic industry . . . Radioactive wastes are a greater potential danger than the fallout of atomic bomb tests."[5]

The academy's concern was not shared by the AEC, which wrote optimistically in its 1959 annual report: "Waste problems have proved completely manageable in the operation of the Commission . . There is no reason to believe that the proliferation of wastes will become a limiting factor in future development of atomic energy for peaceful purposes."[6] Not surprisingly, no coherent policy emerged at that time, and within the next few years, as more commercial reactors came into operation, there was a marked growth of public concern about waste management problems.

One method agreed upon for high-level radioactive waste disposal was burial in a stable geologic medium. In 1971, the first site chosen was a salt mine in Lyons, Kansas. Much to the surprise of the AEC, serious objections to this decision were raised by a Kansas state congressman who brought the issue directly to the public and generated opposition in the Kansas house and senate. He argued that the selection of waste disposal sites had to take into account more than technical feasibility. Any decision, it was argued, had to reflect the views of the people. The federal government alone could not decide whether the salt mine should be used for nuclear waste disposal. In addition, two local scientists from the Kansas Geological Survey maintained that government calculations about the safety of the proposed repository were hasty and inadequate. The AEC, undeterred and still convinced that the public in Kansas would accept their decision, proceeded to carry out the confirmatory tests. However, in the summer of 1971 the AEC learned that a local salt mining company had started digging two or three miles from the proposed site and, to make matters worse, some 175,000 gallons of water had mysteriously disappeared during a solution mining operation in a nearby mine. It did not require extensive calculations to predict that the Lyons site might easily become Lake Lyons. In February of 1972 the AEC withdrew its plans to develop the Lyons site, but by then the public had been awakened to the fallibility of AEC procedures. Resentments generated by this encounter between the AEC and the state helped to create the climate of distrust that has clouded most subsequent attempts to deal with nuclear wastes.

Despite the Lyons experience, public resistance to the AEC siting decisions did not unduly concern the commission. While some notice was paid to nontechnical issues (the AEC reports published at that time carried the usual acknowledgement of the necessity to consider the social and political factors in the decision-making process), the AEC generally ignored these considerations in practice.

Moreover, the strategy of storing wastes irretrievably underground raised a new set of difficult questions:

- What would be the consequences if it was found subsequently that safety, health, or environmental hazards resulted from this disposal method?

- What should be the ethical considerations? For example, in view of the possibility that with advanced technology nuclear wastes might become a source of new energy technologies, did the society that had consumed so much of the world's energy resources have an obligation to leave them in retrievable form?

If the answer was affirmative and the wastes were to be stored in a retrievable mode, then another set of questions was posed:

- Who would guard these wastes, which, in some cases, would remain hazardous for thousands of years?

- What kind of society would be required to maintain perpetual surveillance over radioactive material?

- Would their retrievability increase the likelihood of terrorism and nuclear threats?

The AEC opted for retrievable storage, and in 1974 attempted to develop a Retrievable Surface Storage Facility. This time its environmental impact statement (EIS) received poor ratings from environmentalists, state and local governments, and eventually the Environmental Protection Agency (EPA). The project was abandoned.

By 1975 the AEC (reorganized as ERDA, Energy Research and Development Administration) began to explore another site, this one in Carlsbad, New Mexico. Although originally proposed for permanent storage of transuranic wastes from military programs, the Department of Energy during the early part of the Carter Administration also favored adding a retrievable-storage capacity for 1000 commercial spent fuel rods. In addition, this project, known as WIPP (Waste Isolation Pilot Plant), would have a facility for experiments with high-level defense wastes.

During the lengthy technical preparations for implementing the project, public concerns began to mount locally and nationally. By May of 1978, thirty-three states had passed laws aimed at controlling some aspect of radioactive waste management.[7] Six of these states—Connecticut, Iowa, Maine, New York, Oregon, and Wisconsin—now prohibit construction of new reactors until the disposal issue is settled.[8]

Superimposed on the growing concern about waste storage sites, uneasiness about the transportation of radioactive wastes across states increased as the plans for WIPP took shape. Carlsbad citizens became alarmed by the publication of reports estimated that sixty-five truck and train loads of waste would be transported across the state each week. Furthermore, during the preliminary excavations, substantial deposits of potash, natural gas, and oil were discovered, thereby arousing the indignation of local workers who protested interference with future employment opportunities. Existing federal legislation prohibits waste deposit sites in areas with potential commercial assets. Thus the workers have been abetted by the government on one side and the antinuclear and environmental forces on the other, since environmentalists continue to question the adequacy of salt as a geologic medium. More recently, Native Americans have protested that development of the proposed site would destroy their religious grounds.

In February 1980, five years and $90 million after the project was first adumbrated, President Carter cancelled WIPP, leaving open the option of further investigations for possible future use. The Carter Administration had decided that it would be necessary to test four or five other sites with differing geologic characteristics before agreeing to the construction of a pilot test disposal facility. Three days after the Reagan administration took office, the Department of Energy announced that it would proceed expeditiously with WIPP, and that nothing more was needed from the state, legally or officially. In the case of WIPP, the federal agencies appear once again to have neglected to take serious account of the concerned public in the early stages of policy planning and to provide an adequate technical basis for their decision.

TECHNICAL UNCERTAINTIES

The history of radioactive waste management frequently demonstrates a naivete about the complex political realities associated with the disposal of nuclear wastes. The technical problems of nuclear waste management are difficult as well, and many uncertainties are yet to be resolved. Consequently, the scientific community cannot reach consensus on several of the basic issues. Differing opinions are put forward as to the possibility of accidents if radioactive materials are stored in various geologic formations. On the question of whether the levels set for radiation controls are adequate, one finds competent scientists testifying on both sides. As recently as April 1979, a group of scientists meeting under the aegis of the National Academy of Sciences could not state with much precision how safe is safe.[9] Their report notes the considerable difference of opinion among experts about the human response to a low dose of radiation. It also raises questions about the risks associated with nuclear proliferation and terrorism. The report concludes that these risks cannot be calculated accurately by technological analysis alone.

Another NAS report, written in the spring of 1979, was not widely circulated because "some members of the committee believed the report emphasized one interpretation of the available scientific data to the virtual exclusion of other possible interpretations that had been discussed by the committee."[10] Not until a year later (1980), and then with two members still dissenting and with qualifying reservations from others, was it possible to release the report, which incorporated the two conflicting views of the models to be used in calculating cancer risks.[11]

NAS President Philip Handler, forced to play Solomon, wrote that one model could be used "if social values dictate a conservative approach," while the other should be employed "if one wishes to accept scientists' best judgment, while recognizing that the data simply will not permit definitive conclusions."[12] The report made clear once more that although compromise is now possible, consensus still is not.

President Carter established an Interagency Review Group on Nuclear

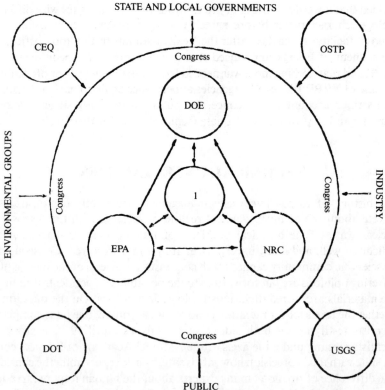

Source: Roger E. Kasperson, "The Dark Side of the Radioactive Waste Problem," in *Progress in Resource Management and Environmental Planning*, ed. T. O'Riordan and R. d'Arge (New York: Wiley, 1980), p. 138.

Figure 6.1   The Nuclear Waste Management System (after Jet Propulsion Laboratory, 1977.)

Waste Management (IRG) in 1977. Chaired by the Department of Energy (DOE), which had prepared an initial draft report, the IRG brought together thirteen executive branch departments and agencies concerned with various aspects of the waste management problem. This new coalition was "to formulate recommendations for establishment of an Administration policy with respect to long-term management of nuclear waste, and supporting programs to implement this policy."[13] (Figure 6.1 demonstrates the current relationships among various agencies and the unresolved question of where the ultimate responsibility lies.) Many citizen concerns were addressed by this group, which consulted experts inside and outside of government. The IRG also submitted successive drafts of its report to more than 200 persons and public interest groups for revision. In addition, high-level officials from the State Department and from other government agencies participated in extensive public hearings in Boston, Denver, and San Francisco.

Despite a number of shortcomings (e.g., inadequate funding for the participation of private citizens who were often contending with highly paid experts from the utilities; insufficient notice of the times of hearings; and some doubt that public testimony had been more than an opportunity for the public to vent hostile feelings) the hearings allowed a thorough exposure of many positions, particularly those at either end of the spectrum—the intensely pro and anti positions. More than 3,000 written comments were received and answered after which the Office of Science and Technology Policy (OSTP) prepared a public document with an analysis of the testimonies. It is generally believed that this specific undertaking represented a major step forward toward resolution of the conflict on a battlefield littered with futile initiatives.[14]

President Carter received the final report of IRG in early 1979, but political considerations delayed his congressional message for a national policy until February 12, 1980. Even then the policy appeared to be one of more planning to plan. The president initiated preparation under DOE leadership of a National Plan for Radioactive Waste Management. At that time he established a State Planning Council (SPC) composed of eight state governors, three state legislators, a mayor, a county official, and the secretaries of energy, interior and transportation, as well as the senior executive of the EPA and a Native American tribal chief. The SPC would work toward the resolution of difficult federal-state-local relations and establish the criteria for reviewing the National Plan.

The President's message also included the dictum: "It is essential that all aspects of the waste management program be conducted with the fullest possible disclosure to, and participation by, the public and the technical community."[15]

To date (July 1981) there has been little, if any, public participation in either the National Plan or the State Planning Council.[16]

THE EMERGENCE OF PUBLIC PARTICIPATION IN THE DECISION-MAKING PROCESS

Running through the history of policymaking in nuclear waste management, there is evidence of a rising tide of expectations and entitlement on the part of the public to having a significant voice in the decision-making process. This in part reflects the growing awareness that much bad judgment and incompetence was masked by military and industrial secrecy. There was a conviction (held particularly by public interest groups) that open debate would have resulted in better and more acceptable policies. The new sense of entitlement also reflects a general tendency toward issue orientation. As the percentage of the public voting in national elections has decreased in the past two decades, it has risen on referendum questions. Party affiliation counts less than the candidates' positions on particular issues. Emotionally charged questions (gun control, busing, abortion, gay rights, the Equal Rights Amendment, and nuclear energy) elicit passionate responses from otherwise apolitical individuals and groups. Ticket-splitting has increased by 25 percent in twenty years, as has the percentage of independent voters who will vote for candidates based on their position on one specific issue.

Environmentalists, although not as well funded or focused as many other single-interest groups, have begun to have a marked political effect. Increasingly their views are taken into account by candidates, who perceive them as an influential force, better educated and more likely to vote than the average citizen. Environmentalists are not necessarily antinuclear (although in the past few years there has been an increase in the coalition of antinuclear and environmentalist groups), but a majority of these groups—Friends of the Earth, the Union of Concerned Scientists, the Sierra Club, the Natural Resources Defense Council, the Environmental Defense Fund, the Consolidated National Intervenors—has been active in initiating legal proceedings where nuclear-related issues appear to threaten the environment.

This dramatic increase of interest in environmental concerns had modest beginnings. During the 1950s, an era marked by an optimism that corresponded to the growth of the economy, little public discussion occurred either about the environment or about nuclear energy (the latter in part because most of it was military and therefore secret). A greater faith in the government marked that earlier period as well; the general who symbolized victory in World War II was elected president. During his tenure, Dwight D. Eisenhower launched the Atoms for Peace program, which was seen as the beginning of a new era, when radioactive isotopes would be used in medicine, agriculture, and engineering. Visions of an unlimited, cheap energy source abounded.

With the publication of Rachel Carson's book *Silent Spring* (1962), concern with environmental issues began to grow. A *New York Times* book review exhorted the public to action: "It is high time for people to know about these rapid changes in their environment, and to take an effective part in the battle that may shape the future of all life on earth."[17]

The new mood and interest of the public was reflected in the rapid succession of laws that began with passage of the National Environmental Policy Act (NEPA) in 1970. Until that time, agency analysis and decision making were open neither to the public nor even to specialists outside the agency. As a result of NEPA, decisions on radioactive waste management were subject to review by independent committees, and environmental impact statements were mandated. Passage of the Freedom of Information Act stimulated increasing scrutiny of government actions by citizens and advocacy groups who now had access to papers and proceedings previously closed to them. AEC misdeeds and misjudgments became public information. At the same time, judicial action made it possible for citizens and public interest groups to take legal action in opposition to decisions that represented dangers to the environment. A landmark decision of the courts, which gave a monumental boost to the morale of nuclear opposition groups in the United States, asserted that "NEPA was not a vague testament of pious generalities but an unambiguous demand for reordering of priorities in specific decision making procedures, including, very particularly, those of the AEC."[18]

THE NEW PUBLIC

As the importance of external and internal reviews came to be recognized, a new cadre of "public" experts was created and drawn into the nuclear-waste-management fray. By and large this "public" grew from an informal network of scientists, engineers, social scientists, lawyers, and policy planners working outside of government and industry. A number of them, acting on behalf of intervenor organizations or as "informed" citizens, participated in the Nuclear Regulatory Commission (NRC) Goals Task Force (1976), an exercise that attempted to incorporate social and institutional factors into its study of technical problems. This was followed by the Chicago Conference (1976) in which five government agencies participated (ERDA, NRC, EPA, Council on Environmental Quality, and National Science Foundation). Although public participation was sought by the agencies, the meetings as described by Metlay and Rochlin[19] were, in retrospect, too formal, too conflict-ridden, and as a result, not successful.

A later effort, the American Physical Society's Report (1977), different from the others, made two distinct contributions. First, although it did not solve the debate, it clarified the issues. Second, it demonstrated that it was possible to identify within the scientific community a group of disinterested, technically competent professionals who would be able to comprehend the technical problems and produce an unbiased report. By seeking the advice of this group as well as that of the more traditional consultants—industrialists, government officials, the wholehearted nuclear supporters within universities, the IRG (although it lost the momentum it had gathered and its consensus may yet wither from a political disease) set a standard for future government planning. For the outsiders, who have since become international consultants, the challenge has been to maintain their once-neutral positions—which is not easy when the long-term assimilation of technical and political issues required to make one an "expert" often also makes one a partisan.

Since 1978 the Keystone Center in Colorado (a nonprofit education organization interested in environmental issues) has sponsored a series of workshops on national radioactive waste management issues. Two workshops examined public participation. Several members of the group were veterans of earlier nuclear waste studies and since the first Keystone meeting have joined other workshops, among them the Harvard-Aspen Institute's Radioactive Waste Management Workshop (1979); Resolve, Center for Environmental Conflict Resolution Workshop (1979-81); and the League of Women Voter's Advisory Board (1980).

Several members of the Keystone group have taken an active part in recent deliberations by meeting with DOE officials (a number of whom attend their meetings), presenting their views to policy planners in Washington, and most recently serving as advisors and observers for the State Planning Council, chaired by Governor Richard Riley of South Carolina.

The League of Women Voters (whose members chaired DOE public hearings) has become actively involved in the Keystone, Aspen, and State Planning Council meetings. During the past several years, the League has launched a vigorous discussion of public participation in nuclear energy issues. Having published *A Nuclear Waste Primer* (1980), it will also publish a comprehensive review of nuclear energy in 1981, and during the same year launch a National Nuclear Energy Education Program. The League's interest was galvanized by the "confusion and contradictions" that surrounded Three Mile Island and by its consequent assessment that "media, government, industry and special interest groups do not, for the most part, provide the balanced, readable information citizens need to understand these issues and to participate in public policy decisions."[20]

Because they are organized on national, state, and local levels, the league's activities will be carried out at the major tension points surrounding the disposal of radioactive wastes. That is, their members will be involved in the areas of struggle for decision-making control between and among federal agencies, state governments, and local citizens.

While the league's stated aim is to educate the public from many different perspectives, few other groups attempt such an evenhanded approach. Increasing polarization marks nuclear discussions as the numbers of avowedly pronuclear and antinuclear groups grow rapidly. In 1975 the *Directory of Nuclear*

Activists recorded 149 antinuclear organizations; by 1980 the numbers exceeded 1,500.[21] The pronuclear groups that are catalogued by the Atomic Industrial Forum (AIF) have more than doubled during the same period, to approximately 200 at present, including regional chapters of national organizations.

Many of these groups differ from more traditional public interest groups such as the league or even antinuclear groups such as Critical Mass in that they are supported primarily by private industry, most often the nuclear and the electric utility industries. In the AIF directory, the addresses of the majority in the category—public participation—are the public relations division of a utility company.[22]

Americans for Energy Independence (AEI), organized in 1975 to achieve rapid energy independence, brings together members of many communities—among them, business, labor, academic, industrial and consumer groups. Dr. Hans Bethe, the distinguished Nobel laureate physicist and staunch supporter of nuclear energy, chairs the board of directors. During the same year AEI came into existence, Dr. Bethe and other scientists in AEI helped create Scientists and Engineers for Secure Energy (SE₂). Organized in response to California's referendum on nuclear power, SE₂ sought to defend the role of science and technology in the nation's future. Unlike AEI, a nonprofit organization that cannot lobby or testify before Congress, SE₂ was set up to do both.

By 1979 they were recruiting international members. The American group, chaired by the former president of Rockefeller University (Frederick Seitz), has, like the Union of Concerned Scientists, many distinguished academic scientists in addition to Bethe (Edward Teller, Eugene Wigner, Robert Hofstadter) among its members. Their goal is to convince the government and the public that nuclear power is safe. To that end they have testified in favor of Windscale's reprocessing plant in the United Kingdom, argued in Congress for the breeder reactor, challenged in court the NRC decision to terminate the GESMO (Generic Environmental Statement on the Use of Recycled Plutonium in Mixed Oxide Fuel in Light Water Cooled Reactors) hearings; sponsored a nuclear waste disposal seminar; and are "considering ways in which to challenge local bans on transportation of nuclear materials and . . . participating in efforts to speed up . . . the Federal decision on reprocessing and final disposal of wastes."[23]

Nuclear Energy Women, an organization assisted in its education effort by AIF, has some fifteen regional chapters, all located in power companies. The members of this organization believe that nuclear power means growth, production, and jobs. In addressing them last year, Dr. Estelle Ramey, an endocrinologist on the faculty of The George Washington University Medical School, attributed women's advancement in the United States to the availability of cheap and abundant energy. "Women," she added, "needed it to continue to be liberated. If that meant nuclear energy, so be it."[24] Margaret Bush

Wilson, chair and legal counsel for the NAACP and board member of the Electric Power Research Institute (EPRI), has strongly stated the same argument for black labor, male and female.

Labor unions have been regarded traditionally as strongly pronuclear; the AIF directory records nineteen pronuclear unions. However, Logan and Nelkin have reported that the nuclear issue has become a source of conflict within and between labor unions.[25] For example, the United Mine Workers (UMW), like their counterparts in Great Britain, concerned with strengthening their own industry, have been opposed to nuclear energy for a quarter of a century. As might be expected, the AFL-CIO, with a large membership of construction unions, has strongly supported nuclear power despite dissension in its ranks. Were WIPP to be built in Carlsbad, New Mexico, it would provide long-term employment for more than 300 construction workers in a city of 28,000 people, where, as Logan points out, its major industry, potash mining, is declining. The addition of jobs would make a significant difference.[26]

In Massachusetts, a state where its senior senator (Kennedy), the lieutenant governor (O'Neill), and a representative (Markey) have lobbied for a moratorium on nuclear power plant building, a coalition of Citizens for Participation in Political Action (CPPA) and the local Clamshell Alliance introduced a number of confusing nonbinding antinuclear referendums on the November 1980 ballot in one-third of the state's districts. The antinuclear initiatives won by a narrow margin, but more interesting, and typical of the confusion surrounding nuclear initiatives, were the results of those voted on in the precincts surrounding the Pilgrim nuclear facility in Plymouth, Massachusetts. There were two ballots; one opposed future construction; the other supported it. Both were defeated. The Boston Globe quipped: "A resounding 'maybe' in Plymouth."[27]

Another antinuclear referendum (the first since TMI), which advocated a complete shutdown of nuclear power in the state, was defeated by 70,000 votes in Maine earlier in the year. For the pros, it was an unmistakable victory; for the antis, it was interpreted as a mistake in wording the referendum. Rather than ask for a complete shutdown, they believe (on hindsight) that they should have tried to achieve an orderly phasing out of nuclear power.

Only twenty-three of the nation's fifty states have provisions that allow independent initiatives to be placed on ballots. In November, 1980, South Dakota, Oregon, Washington, Missouri, and Montana had antinuclear initiatives on their ballots, while five other initiatives were being prepared for the next election.[28] Most of the referenda were modeled on an earlier Montana referendum (1979), which bans the further construction of nuclear power plants until adequate means of waste disposal are found; the recent Montana referendum was to ban radioactive waste disposal. The pronuclear vote prevailed in Missouri and South Dakota; the antinuclear, in Washington and Oregon; while that in Montana was so close that each side claimed victory for more than a month. The final count—172,909 (in favor of banning disposal of nonmedical radioactive waste materials) to 172,493 dramatically demonstrates the closeness of opinions regarding nuclear waste management. Whether the states have the right to enact these laws will likely be determined by the courts, which in 1981 invalidated the Washington State initiative banning out-of-state wastes but upheld the California statute stopping new power plants until a waste disposal system exists.

Antinuclear activities have also been abetted in the past four years by an informal group of more than fifty foundations and philanthropists whose latest conference, the Alternatives to Nuclear Conference, intended to show that alternative energy sources could be found that are technically, socially, and economically sounder than nuclear power. The group's activities range from supporting the travel of scientists who have testified at nuclear waste hearings in Great Britain and Germany, to sponsoring rock-and-roll musicians. A concert in Madison Square Garden (1979), a recording, and a movie of the event, yielded more than $1,000,000. These funds were used to set up another nonprofit foundation, Musicians For Safe Energy (MUSE), which has distributed the money to grass-roots antinuclear organizations throughout the United States. Many of the MUSE musicians are West Coast artists, veterans of antinuclear campaigns that began with the unsuccessful effort to pass Proposition 15 (1976), which proposed that no more nuclear plants be built in California until the federal government certified a way of disposing of radioactive waste.

The past decade in the United States has been marked by the rapid growth of volunteer and professional organizations that deal with controversial issues at local, state, and federal levels; nuclear energy and, particularly, nuclear waste management have been for many their major concerns. As the number of activities has increased and the issues themselves become more complex politically (and sometimes technically), the polarization of the positions has become more pronounced and the moderate position has become increasingly difficult to identify. This phenomenon is not uniquely American; to varying degrees, the struggle is international.

NATIONAL DECISION MAKING AND INTERNATIONAL REPERCUSSIONS

As Three Mile Island demonstrated and the London *Economist* headlined, "We All Live in Harrisburg." Even when mishaps are minor, an ever-vigilant news industry, sensitive to the public's ambivalence about nuclear energy (created in part by the potential for catastrophic destruction), rushes in with elaborate coverage. Nuclear power problems are always newsworthy.

A protest in France can be relayed live around the world, while a scheduled demonstration in Germany can draw antinuclear forces from most of Europe in less than a day. As in the United States, it is not only the antinuclear activists who have predominated in nuclear wastes debates. Before TMI occurred, when nuclear waste concerns—whether proliferation or disposal—dominated nuclear fuel cycle problems, an intra-European cadre of nuclear critics joined their American counterparts and shuttled between American and European courtrooms, government offices, and previously little-known cities in order to testify for or against specific nuclear waste disposal plans.

When TMI occurred, many members of this group were participating at the Gorleben hearings in Germany, while in Sweden, where they had also served as government consultants, a recommendation regarding high-level waste was just about to be introduced. TMI forced a parliamentary decision to proceed with a national referendum rather than any further plan for nuclear energy. The hearings also resulted in the dismantling of plans for a major reprocessing and waste-disposal center.

As consultants to governments or "friends of the court," the international nuclear critics differ from the antinuclear movement as a whole in a significant manner; they are technically expert. An informal sociometry reveals a high percentage of physicists, biologists, and mathematicians. Within this group, attitudes toward nuclear power range from avowed antinuclear (Amory Lovins, Henry Kendall) to pronuclear but concerned about specific technical issues of waste management and safety (Frank Von Hippel, Walter Patterson, Albert Wohlstedter). The American group (Irwin Bupp, David Deese, Nancy Abrams, John Holdren, Terry Lash, Joel Primack, Gene Rochlin, Roger Kasperson, Todd LaPorte, Daniel Metlay, et al.) has many European counterparts (Mans Lonnroth, Meyer-Abich, Thomas Johansson, Thomas Conroy, Jean-Claude Derian) who have been instrumental in defining the range of issues to be addressed in nuclear waste legislation. Because they are not inside the (nuclear) system, they act as "public" intervenors. In Sweden (KBS nuclear fuel safety report), Germany (Gorleben) and the United Kingdom (Windscale), they have sharpened the debate to make explicit the technical, social, political, and economic issues of nuclear waste management and the interrelatedness of these variables.

From a pronuclear perspective, the group is seen as uniformly antinuclear, while the differences within the group are, in fact, significant. For example, Friends of the Earth (FOE), with headquarters in London and offices in San Francisco and Paris (Les Amis de la Terre) represents to many in the pronuclear establishment a "well-respected leadership" in the environmental movement. To others, such as Sir Fred Hoyle, a distinguished astronomer, they are viewed as a "Marxist group with . . . Kremlin-inspired instructions" to develop an antinuclear movement in the West, thereby destroying its energy supply. This in turn would bring about a Russian "victory in the world struggle."[29] That view was supported by the former chief scientific advisor to the British government,

Sir Alan Cottrell, who in the introduction to Hoyle's book expounded its "refreshing common sense" in contrast to "the hysteria of the anti-nuclear environmentalists.

Friends of the Earth cried "foul," and each copy of Hoyle's book carried an apology from the author and the publishers. Litigation was avoided, but the public for whom *Energy or Extinction* was written undoubtedly learned as much about the polarization between the environmentalists and well-known members of the British scientific establishment as it did about the need for nuclear energy.

FOE continues to evolve organizationally. It began as pro-environment with room for both antinuclear members (Lovins) and nuclear critics of "evaluators"[30] (Patterson). During the past five years the nuclear critics have been displaced by antinuclear supporters (Conroy, Flood), while splits have developed between the no-growth factions and the growth-but-not-through-nuclear members.

The opposition to nuclear power within FOE was exacerbated by the results of the first large-scale public examination of nuclear-fuel reprocessing (the Windscale Inquiry), begun in 1977.

FOE's membership surged during the 100 days of the hearings. Because of the legalistic format, those who testified on behalf of FOE took a moderate approach and appeared to narrow the distance between themselves and the nuclear establishment. This, in turn, created increasing criticism of its leadership by FOE's rank and file, who resented the emergence of a new "autocratic [environmental] elite."[31]

As a result of the organization's internal struggle, FOE's most recent campaign is concertedly antinuclear. With a budget of one million pounds, they aim to make nuclear energy a major issue of the 1984 general election in Great Britain. The first four years of a five-year strategy will be devoted to increasing public awareness through education, summer schools on nuclear power, and community organization. The final year will be devoted to the support of antinuclear political candidates. Throughout the campaign, the problems of waste disposal will be discussed in the context of the morality of risking nuclear power development in the absence of adequate disposal systems.

For many in the United Kingdom, the Windscale hearings were a model of public participation. Lord Brian Flowers, author of the influential White Paper *(Nuclear Power and the Environment)*[32] that addressed the environmental issues of radioactive-waste management and nuclear proliferation, reported: "Although the Windscale report should have stated the opposition position more clearly, it was, nevertheless, a magnificent attempt.[33]

Lord Flowers now believes that the environmentalists have not acknowledged the actual gains made in nuclear waste management and safety problems since Windscale. Observing that vitrification is under "pretty good" development and that the safety of the fast breeder has increased in the last five years, he regrets that a "vigorous program" on nuclear waste is being held back by environmentalists. For him, the challenge now is to educate the public about the need for commercial nuclear power; its development should not be impeded by the "red herring" issue of the threat to civil liberties were the world to become increasingly nuclear.

Despite Windscale, open public debate that can affect policy decisions is often difficult to achieve in Great Britain, in part because of the structure of the courts and government, and in part because secrecy is endemic in the British system. For example, in the first public inquiry into test drilling for the possible disposal of high-level radioactive waste (Ayr, Scotland, February 1980), the government forbade testimony on nuclear waste dumping and alternative methods of disposal. The Scottish Conservation Society organized its own group (the People's Planning Inquiry Commission) and held parallel meetings of which 99 percent of those present were opposed to the drilling proposals.[34]

Nuclear critics believe there is little likelihood that public intervention will be significant in Great Britain or that it will have an effect on American or other European public programs for nuclear waste management.[35] Beyond this, nuclear activists, both pro and con, agree that the British public is bored with nuclear issues. "In fact," said a nuclear observer, "a nuclear debate is a luxury. We have so many more horrendous problems."

SWEDEN: A SPECIAL CASE[36]

If any public is entitled to be bored by the nuclear energy debate, it is the Swedes. Since 1976 when the nuclear issue figured prominently in the defeat of the Social Democratic Party, which had been in power forty-three years, nuclear energy has played a major role in Sweden's political debates. Until that time, the country had a history of strong support for nuclear energy. Ironically, it was the environmentalists who had spearheaded the drive to nuclear power, when in the 1950s, in order to preserve the country's rivers, they opposed increasing the development of hydropower. Accordingly, in the early 1960s, the state-owned utility chose nuclear power, and it received strong political support for almost twenty years.

Sweden is an affluent, highly industrialized country with a homogeneous, well-educated public, an enviable standard of living, and a record of public welfare programs that dates back to 1911. Unravaged by war, they had the technical capability to develop the most advanced nuclear industry in Europe and looked forward to a thriving nuclear export business. Initially, in 1967, siting was the only public issue because it was assumed that nuclear waste would be disposed abroad. The issue of reprocessing changed the balance, and the first discussions of nuclear waste and safety in the Swedish Parliament marked the beginning of the splits within the government. Social Democrats, with their strong pronuclear program, were attacked from the left and right, (that is by the Communists and by the Center Party, which had its roots in the Agrarian Movement). They lost to a coalition of three nonsocialist parties, one strongly antinuclear (Center Party), one mildly pro (Liberals), and one strongly pro (Conservatives). Hannes Alfven, a Nobel Prize winner in physics, once pronuclear but later strongly antinuclear, was credited with helping to bring about the Center Party's victory by campaigning for its leader, Thorbjorn Falldin.

Before their downfall, the Social Democrats had launched an ambitious nuclear energy public-education program in an attempt to win over what they saw as an ambivalent public. More than 8,000 study groups were organized under the aegis of labor organizations throughout the country, and discussions were held to convince the public of the worthiness of the nuclear option. Those group leaders who were wholeheartedly pronuclear stressed energy needs; the less certain stressed nuclear risks. To date it has been difficult to estimate the result of this public education program. It appears that those who were antinuclear did not change their position; that those who were skeptical had their position reinforced and in many instances became more skeptical; and that those who were pronuclear remained so.[37] The percentage of people who could make no decision regarding nuclear increased after the study group by 7 percent. As the public began to learn more about the risks, their attitudes hardened, thereby convincing a number of Social Democrats that public education had hindered rather than helped them achieve their goals.

Later, a study of how the Swedish population learned about nuclear energy revealed that 79 percent received their information from TV; 47 percent believed that the most trustworthy information it received came from broadcasting and TV sources. Only 2 percent cited the study circles, which had been such a large part of the public education effort. Obviously, the final word on public education has not been demonstrated.

The international nuclear critics took a more active role in the next public nuclear issue. Following the defeat of the pronuclear Social Democrats (1976), the ambivalently antinuclear coalition government submitted a proposal for the disposal of high-level wastes from reprocessing (KBS report, 1978) to an international review committee (Rydberg-Winchester Review). The dissent that followed release of the committee report and the general dissatisfaction with the coalition's energy policies created a political controversy so divisive that the coalition government collapsed. The new minority leaders, the Liberals, planned to initiate a bill that would support the fueling of two reactors already built and would also recommend a cautious expansion of nuclear power. Three Mile Island brought an abrupt halt to these plans.

Concerned that they would lose the next election, the government turned back to the public the question of whether nuclear power should be continued. A national referendum was proposed. The public was divided on the political wisdom of the proposal for a national referendum; only three have been held previously and were not constitutionally binding. This time, all parliamentary parties agreed to abide by the outcome. Some of the public perceived the referendum as a failure of government to make decisions while others welcomed it as a chance to test public opinion. By 1980 the Swedish public, which had been bombarded by information for more than a decade, was undoubtedly the most well-informed public in the world about nuclear energy.

The ongoing public debate that led up to the referendum began in a concerted fashion in 1969 when students at Gothenburg University protested against the choice of a site for a reprocessing plant which they felt would create health hazards and problems about plutonium, and would also destroy local beaches.[38] As in the United States, the earlier debates were isolated and intense (Linus Pauling, an American Nobel chemist, testified against the first major nuclear project in Sweden in the early 1960s), but the impact was local, not national. By 1973, opposition had mounted and antinuclear books had begun to appear. Friends of the Earth, with headquarters in London but with a mobile membership, collaborated with local Swedish citizens. The fight was on.

An environmental movement, led by Bjorn Gillberg, dedicated itself to improving Sweden's quality of life. In 1974 a survey reported that the Swedish public had more faith in Gillberg's ability to solve environmental problems than it did in either the consumer ombudsman or the Office of Consumer Affairs. By 1975, Gillberg's organization split internally because of many of the same ambiguities that beset public intervenor groups: Are the goals to change the government, halt nuclear energy, or make it safe? By the time the organization had lost its influence, however, it had helped to develop a large-scale public awareness of the nuclear industry.[39]

By 1978 the People's Campaign Against Nuclear Power had collected several hundred thousand names requesting a referendum on nuclear power.[40]

Three Mile Island, as mentioned earlier, was the traumatic event that forced even the Social Democrats to relinquish their long-held pronuclear position and to join with the coalition government to propose a referendum to be held *after* the general election in September 1979. It also captivated the Swedish press and TV. Many programs were preempted for hours of debate on the pros and cons of nuclear energy, and many of the international nuclear activists were summoned to participate.

The referendum, which came finally after 10 years of public debate, offered three choices, none of which alone captured a majority of the public's consent:

- *Alternative I.* Supported by the Moderate Party and business organizations (pronuclear). This proposed that all twelve reactors—those already operating as well as those under construction and those in the planning stage—be used until ready to be decommissioned in approximately twenty-five years. This alternative stressed Sweden's dependence on foreign oil (70 percent) and the need for industrial expansion to ensure full employment.[41] (18.9 percent in favor)

- *Alternative II.* The Social Democrats and the Liberals also proposed twelve reactors and advocated phasing them out by the year 2010. In addition, on the back of the ballot, they stressed the need to move away from electricity intensive industries and housing, to increasing the use of renewable resources and to turn the nuclear industry from private to public ownership. Conservation would be vigorously promoted. (39.1 percent in favor)

- *Alternative III.* The Center and Communist Parties requested a phasing out of the six reactors currently in operation within ten years. No others would be completed or built. Conservation would be stressed and there would be a substantially increased investment in renewable sources of energy. (38.7 percent in favor)

Approximately 75 percent of the public voted—15 percent less than in general elections. The long-range political results are unclear. For many, the referendum was a welcome end to a long, tedious debate. For others, the struggle would continue.

What is clear from the referendum results is that they have had an impact on the international perceptions of nuclear power. The international press and national governments have interpreted the results according to their own predilections. A sampling of headlines reveals the disparity:

SWEDISH VOTE ENCOURAGES OTHER PRONUCLEAR EUROPEAN EFFORTS

The New York Times, 3/30/80

AN OVERWHELMING MAJORITY OF SWEDES VOTED AGAINST ANY FURTHER EXTENSION OF NUCLEAR POWER IN A SPECIAL ADVISORY REFERENDUM LAST SUNDAY.

Nature, 3/27/80

SWEDEN APPEARS TO ENDORSE NUCLEAR ENERGY

International Herald Tribune, 3/24/80

SWEDEN'S NUCLEAR REFERENDUM WOUND UP WITH EVERYONE CLAIMING A VICTORY.

Nucleonics Week, 3/27/80

A Swedish nuclear authority who stated that Sweden had voted to phase out nuclear power was surprised indeed when presented with the above headlines.[42]

The environmental movement has continued to grow in Sweden, but it is not deeply troubling to the government, in which many officials are themselves environmentalists. Far more distressing to the government is the galloping growth of public criticism of the bureaucracy and the scientists and engineers within it. For the first time in their history, scientists have come under attack because they are no longer seen as disinterested by the public and antinuclear groups. Interviews with Swedish officials reveal the depth of their despair at this change in the public's perception of the entire scientific profession; it is a marked discontinuity from the apolitical past of science in Sweden. In yet one more country, nuclear power has become the symbol of widespread discontent with big government. This discontent and the desire by the public for greater access to the decision-making process brought Sweden to the unique referendum of March 1980. And as the headlines demonstrate, the repercussions have been felt far beyond its borders.

At the present time, the Swedish government is conducting research to examine the history of nuclear power and the public during the past quarter century. As in the United States, many Swedish scientists and government officials interviewed for this project express the belief that television and the newspapers have been biased in their presentation since the very beginning of nuclear power as a public issue.[43]

The Austrian government has been cheered by the Swedish referendum results. Having been surprised by the results of its own nuclear referendum (1978), in which by a narrow margin the country voted not to open its one completed plant, the government has begun once more to test public opinion about its present attitudes. This time the government is abetted by a citizen's initiative sponsored by the nuclear lobby, which is attempting to reintroduce the referendum. The government is proceeding cautiously, as its previous effort to educate the public about nuclear energy gave the antinuclear groups the prominence that allowed them to defeat the referendum. The chancellor, Bruno Kreisky, who threatened to resign if the referendum were defeated, did not. He, too, is no longer issuing strong statements.

At the same time as the government has been preparing a new referendum, antinuclear demonstrators in the western Austrian province of Vorarlberg (near Switzerland) have been publicizing the likelihood that radiation from a nuclear power plant could cross the border between the two countries. The Austrian critics helped Swiss antinuclear forces postpone the construction of the plant. In December 1980 the Swiss federal energy commission, unable to agree whether the country faced an energy shortage, gave a victory to the Swiss antinuclear forces by bringing about a de facto moratorium on new orders for nuclear power stations. As in other countries where open debate shaped the argument, independent scientists testified on both sides of the issue.

Another border neighbor, Czechoslovakia, has been less cooperative; several additional plants are being constructed.[44]

Both the government and industry hope for a "Swedish result," that is, to operate those plants already built or planned, so that its now idle nuclear power plant at Zwentendorf can move into production.

The French, too, were relieved that a democratic debate in Sweden produced the same results that they had achieved by a fiat. Unlike Germany and Sweden, France is without a nuclear safety law or an independent nuclear safety control group. The international nuclear critics who have played such an important role in the U.S., Swedish, and German deliberations have had little impact on French nuclear policies.

Public participation in France provides strong contrast with Sweden. Not only has there been no moratorium or public debate, but quite the opposite: France has plunged into an ambitious nuclear power program that aims to produce 60 percent of the country's electricity needs by 1985, a 46 percent increase in current nuclear output. By 1990 the French aim to generate 30 percent of their total energy from nuclear power rather than the 5 percent it contributes today.

The official attitude—that "there is nothing to be gained for real informing of the public (sic) by holding controversial debates" and "Three Mile Island could not happen here"—has been accepted by a majority of the people, particulary in the light of the cutback of Iranian oil exports and the increase in oil prices.[45]

In June of 1979, when a group of West German antinuclear activists attempted to cross the French border near Cattenom to protest the construction of the forty-eighth nuclear power reactor, they were turned back by French police, and the incident was reported only in the back pages of the newspaper. That the German police had given a list of names of known environmentalists and antinuclear activists to the French gendarmes received no notice (personal communication).

How these actions could be tolerated was explained by the French Ambassador to the United States in 1979.[46] The behavior was, he explained, a reflection of French pragmatism. The French are determined to be first among the nuclear elite. This long-standing goal has been sharpened by their 75 percent dependence on imported energy, as opposed to 55 percent in Western Europe as a whole. In addition, nuclear plants, which are clean, are more attractive to the French than the coal plants, which for decades have contributed to the griminess of many industrial cities. He added that the "speed-up" nuclear decision had been made reluctantly but resolutely.

After a brief flurry of antinuclear activity, which peaked in 1977 at Creys-Maleville, where a protestor was killed by local police, the small antinuclear Ecologist Party began to lose its momentum. The remaining antinuclear groups—CRILAN (Regional Information Committee for the Anti-Nuclear

Struggle) and CCPAH (Coordinating Committee for Protest against La Hague)—have had little success in rousing substantial public support. Their attempt to interfere with the delivery of the first batch of Japanese spent fuel destined for the reprocessing plant at Cogema was dashed by the administration's decision to reroute the ship to England. The protesters had hoped that even if their efforts failed, the recent ruling of a small town in Normandy to prohibit the transportation of radioactive material through its territory would accomplish what the antinuclear forces did not. However, the central strength of the government is such that any local ordinance can be overruled by the prefect, who represents the federal government, on grounds of national interest. In the 1980s, some decentralization of power could occur if President Mitterand weakens the *préfets*.

The chief spokesman for the French Atomic Energy Commissariat claimed, "We are the only country with a real consensus for nuclear energy."[47] He reported that 62 percent of the public favored nuclear energy (May 1979); *Le Figaro,* in a poll conducted at about the same time, reported 47 percent in favor. Just which figure is more accurate is relatively unimportant, because as TMI demonstrated, nuclear attitudes are extremely volatile; one accident in France, and the percentages would shift.

The French, by and large, are a cynical lot, as shown by the results of an opinion poll reported by antinuclear Confederation Française Democratique du Travail labor union officials at the International Scientific Energy Meetings in Nice, November 1979. Eighty percent of the respondents stated that had Three Mile Island occurred in France, the public would never have been told; 40 percent thought that a similar accident probably had already occurred.

Perhaps it is not cynicism, but rather a response to reality. The reins of power are held very tightly by a centralized and authoritarian administrative system. Decisions are made by a small number of top-level government officials protected from external scrutiny. As in the United States, the commercial nuclear enterprise grew out of the military establishment, where secrecy has been the hallmark. However, it is also the standard way of operating in France. A concerted antinuclear movement would be, in effect, a revolutionary statement.[48]

In addition, the decisions to increase nuclear production have been supported by and large by all the political parties, which are unqualifiedly pronuclear. There are few opportunities in the closed political-industrial-military system to challenge the major decisions effectively through the courts, which have sided consistently with the administration. In each of many nuclear-related decisions, the courts have judged the public interest as identical with government goals.[49]

At present, antinuclear activity has been effectively contained. The prominent antinuclear voice is that of the socialist trade union, the CFDT, which, though pronuclear because of employment opportunities offered by the nuclear power plant construction, is concerned that the health hazards to which their workers are exposed are not being considered seriously enough in the planning and operations of the plants. The Communist Trade Union, the CGT, shares these concerns.

It is too early to assess the price to be paid or the advantages to be gained in this race to hold the number one nuclear position. A major accident could swing the pendulum in one direction; a strengthened economy and undisputed leadership in nuclear technology could swing the pendulum in the other. France has taken a well-calculated but significant risk and has become, in effect, an experimental laboratory for rapid nuclear power development. The many countries whose governments and nuclear industries are beleagured by public interest group litigation and citizen demands for participation in decision making and whose legal systems are not closely tied to administration positions are obviously envious. When the doors open, however, whether the lady or the tiger will appear is a moot question. France is carrying out a national experiment, the findings of which—both technical and social political—are likely to affect the long-range development of commercial nuclear energy throughout the world.

MORAL AND ETHICAL CONSIDERATIONS

The long delay in dealing with nuclear wastes has created broad technical and political arguments which raise moral and ethical issues that will prove even more troublesome in the long run. Public participation becomes all-important, since moral and ethical priorities derive from values held by a society. To arrive at decisions that reflect some measure of consensus, these values must be elucidated and given priority.[50]

If, in years to come, there is less concern about future generations than there is today, one set of decisions can be made. If a certain percentage of fatalities is acceptable for workers, other decisions can be made. However, each assumption must be made as explicit as possible so that both the risks and the equity of those risks are fully understood. A program of economic or other incentives should be developed to compensate those individuals and communities who are asked to assume some of the risks connected with radioactive waste disposal. Fairness, a value not often pursued in the rush to industrialization and affluence, will have to be given new attention.

Nuclear power and nuclear wastes are unique in the fears that they arouse. Pronuclear scientists often do not understand this concern; indeed, they express surprise that the public is so concerned about safety when so few people have been killed in nuclear accidents. They point out that although more than 100,000 miners have been killed in coal-mining accidents since the turn of the century (1,300 since World War II), this record has provoked little public outcry. In part, the difference in public concern results from the recognition that mining has killed only miners, whereas the release of radiation from nuclear accidents is likely to harm many others not directly involved in nuclear power generation. The same argument can be used with airplane and automobile fatalities: the majority involve individuals who have chosen the risk, whereas radiation like the less lethal but more extensive effluents from coal-fired power plants is socially imposed and perceived as a restriction on individual freedom by those who oppose it.

In addition, radioactivity—though an accepted part of medical treatment—is popularly associated with destruction. The nuclear holocausts of 1945, the secrecy of AEC activities, and the deceptions about radioactive fallout during the above-ground nuclear tests in the 1950s taint nuclear power's present image. The continued escalation of nuclear weapons manufacture maintains for nuclear power its lethal reputation. And, not insignificantly, fears about that which is invisible but capable of penetrating people, buildings, and the ecosphere may well contribute to the public's wariness of nuclear power.

OVERVIEW

What then can be gleaned from these wide-ranging observations about the public, nuclear energy, and international interdependence?

First, it is clear that the public as a classification has to be more carefully defined in all future planning. As FOE and SE$_2$ are part of the public, so are those groups in whose backyards nuclear wastes will be stored or those citizens who are ignorant of the issues but whose support is needed in order to forge an informed consensus. Public education then, has many levels and different goals.

Second, because of the increasing polarization within the "public" scientific community, its credibility is likely to diminish. During the public debates preceding the Swedish referendum, the newspapers observed that each side was lining up its doctors and scientists. The chairman of the Royal Institute of Engineering said that for the first time, the public was beginning to think that scientists could be bought.

As a part of public education, it has become increasingly important to air the differences within the scientific and science policy community and to make clear how compromise can be achieved.

The totally pronuclear and totally antinuclear groups are unlikely to contribute to a workable resolution of nuclear waste problems. But groups and individuals are arrayed—as in a Gaussian distribution—along a spectrum of attitudes. Those grouped closer to the center have the potential to work together to depolarize the issue so that negotiations can proceed. Those scientists who have been branded antinuclear by the intractables, but are in fact pronuclear yet concerned about a specific aspect of the fuel cycle, have already begun to work with citizen's groups. Environmentalists, fearful of the effects of nuclear energy on the biosphere, have initiated comparative studies of coal and nuclear power. These efforts can be strengthened through cooperation with government and industrial officials.

It will be necessary to take a long-range view. The overselling of the nuclear option has proven to be a hindrance. Only through open negotiation among governments, private industry and responsible groups will polarization be reduced. There will be controversy as the relative risks of nuclear and coal power are explored. Indeed, a major aspect of public education will be in learning the answers to the question, "If not nuclear, what?"

By now it should be obvious that no one group holds title to all knowledge and wisdom. As the public will have to gain much of its information to make decisions about resources and risks from the experts, so will the experts have to listen more carefully to the questions being raised by an increasingly sophisticated public, which has learned to take seriously their rights and responsibilities in a participatory democracy. Three Mile Island, Gorleben, and the Swedish referendum have demonstrated the international interdependence of nuclear issues. The title of Sweden's recent nuclear legislation might well serve as a guideline for the decade ahead: *Time for Reflection.*[51]

NOTES

1. Dorothy S. Zinberg, "The Public and Nuclear Waste Management," *Bulletin of the Atomic Scientists,* January 1979, pp. 134-139.

2. See Eugene G. Skolnikoff, "Implementation: Action or Stalemate?" Conference Paper on Public Issues in Nuclear Waste Management, 1976 (unpublished).

3. For a detailed history of radioactive waste management policy on which this section is based, see Richard G. Hewlett, Chief Historian, Executive Secretariat, U.S. Department of Energy, "Federal Policy for the Disposal of Highly Radioactive Wastes from Commercial Nuclear Power Plants: An Historical Analysis," March 9, 1978 (unpublished); see also: Daniel S. Metlay, "History and Interpretation of Radioactive Waste Management in the United States," in *Essays on Issues Relevant to the Regulation of Radioactive Waste Management* (Washington, D.C.: Nuclear Regulatory Commission, Office of Nuclear Material Safety and Safeguards, 1978).

4. Carroll L. Wilson, "Nuclear Energy: What Went Wrong," *Bulletin of Atomic Scientists,* vol. 35, no. 6 (June 1979): 13-18.

5. National Academy of Sciences, National Research Council. *The Disposal of Radioactive Wastes on Land,* Pub. No. 519 (1957).

6. Quoted in *The New York Times,* July 9, 1979, in David Burnham, "Growing Waste Problem Threatens Nuclear Future," p. 1.

7. Roger E. Kasperson, Center for Technology, Environment and Development, Clark University, "The Dark Side of the Radioactive Waste Problem," in *Progress in Resource Management and Environmental Planning,* vol. II ed. T. O'Riordan and R. d'Arge (John Wiley & Sons, Ltd.: 1980).

8. Similar constraints in California and Minnesota have been reversed recently in the Federal courts, but the final decision as to whether the federal government or the state has the authority to regulate nuclear power issues has not been made.

9. National Academy of Sciences (Steering Committee for the Nuclear Risk Survey, Committee on Science and Public Policy), *Risks Associated with Nuclear Power: A Critical Review of the Literature* (1979), summary and synthesis chapter.

10. Gerald S. Schatz, "Uncertainties in Estimating Effects of Radiation at Low Doses," *News Report* of the National Academy of Sciences, National Academy of Engineering, Institute of Medicine and National Research Council, vol. XXX, no. 10 (October 1980).

11. For a discussion of the two models proposed see: "The Effects on Populations of Exposure to Low Levels of Ionizing Radiation." Committee on the Biological Effects of Ionizing Radiations; Division of Medical Sciences, Assembly of Life Sciences, National Research Council (National Academy of Sciences, 1980, publication pending).

12. Ibid.

13. Memorandum for the Secretary of State et al from President Carter, March 13, 1978.

14. For a detailed history of the IRG report see chapter one in this volume.

15. President's Message to Congress, February 12, 1980.

16. For a discussion of public participation in the National Plan and the State Planning Council see: *Public Participation in Developing National Plans for Radioactive Waste Management,* October 1980, The Keystone Center, P.O. Box 38, Keystone, Colorado.

17. *The New York Times Book Review,* September 23, 1962, p. 1.

18. Walter C. Patterson, *Nuclear Power* (England, Penguin Books: 1976), p. 193.

19. Daniel S. Metlay, and Gene I. Rochlin, "Radioactive Waste Management in the United States: An Interpretive History of Efforts to Gain Wider Social Consensus" (paper prepared for RESOLVE Nuclear Waste Management Process Review Workshop, Palo Alto, CA, Dec. 1979).

20. Personal communication for Nuclear Information and Resource Service.

21. The Directory of Nuclear Activists, P.O. Box 545, La Veta, Colorado 81055.

22. INFO Atomic Industrial Forum, Public Affairs and Information Program.

23. Brochure from SE$_2$, 570 Seventh Ave., Suite 1007, N.Y., N.Y. 10018. More recently they have been joined in their efforts by the Washington based Committee for Energy Awareness, a group composed of scientists on loan from utilities.

24. "Women Gather to Hear Nuclear Power Promoted," by Karen Dewitt, *The New York Times,* Oct. 19, 1979, page A20.

25. Rebecca Logan and Dorothy Nelkin, "Labor and Nuclear Power," *Environment,* vol. 22, no. 2, (March 1980): 6-13.

26. Ibid., Logan.

27. *Boston Globe* November 6, 1980

28. *Energy Upbeat,* vol. 1, no. 1 (April 1980). Published by the Committee for Energy Awareness.

29. Fred Hoyle, *Energy or Extinction? The Case for Nuclear Energy.* (London: Heinemann Ltd., 1977), chapter 1.

30. Term used by William Walker.

31. Brian Wynne, "Decision Making and the U.K. Nuclear Debate—Where do we go from here?" (University of Lancaster, U.K., June 1978, unpublished: and Brian Wynne, "Windscale: A Case History in the Political Art of Muddling Through," *Progress in Resource Management and Environmental Planning,* vol. 2, ed. T. O'Riordan and K. Turner. (New York: John Wiley & Sons Ltd., 1980), pp. 165-204.

32. Lord Brian Flowers, Her Majesty's Stationery Office, (May 1977).

33. Personal communication, May, 1980.

34. Ted Stevens, "Split Over Admissible Evidence at Test Drilling Inquiry," *Nature,* vols 2 & 3 (February 23, 1980): 805.

35. Walter C. Patterson, "Nuclear Waste Management in the United Kingdom, Issues and Options," (Working Paper for Keystone Meeting VI, September 14-17, 1979).

36. For a lengthy discussion of nuclear waste in Sweden see Mans Lonnroth, "The Back-end of

the Nuclear Fuel Cycle in Sweden" (paper prepared for the Keystone Conference on Nuclear Waste Management, 1979).

37. Dorothy Nelkin, *Technological Decisions and Democracy: European Experiments in Public Participation* (Beverly Hills, CA: Sage Publications, 1977), p. 105.

38. Ibid.

39. Mans Lonnroth "Swedish Energy Policy: Technology in the Political Process" in *The Energy Syndrome,* edited by Leon N. Lindberg, (Lexington, MA, Lexington Books, 1977).

40. Sten G. Sandstrom, "After the Referendum," *ATOM* 285 (July 1980), p. 178.

41. For a comprehensive overview see Per Ragnarson, "Before and After: The Swedish Referendum on Nuclear Power," *Political Life in Sweden,* no. 5 (September 1980). Published by Swedish Information Service, 825 Third Avenue, N.Y., N.Y. 10022.

42. Other Swedish authorities also believed that the vote was to end nuclear power. See Ragnarson, "Before and After," section heading "Yes to Phasing Out of Nuclear Power."

43. Interviews with members of Swedish Nuclear Power Inspectorate, September 1979.

44. "Plan by Prague for Atom Sites Worries Vienna," *New York Times,* September 2, 1980.

45. Interview with French nuclear official, September 1979.

46. Ambassador Francois de Laboulaye, in a lecture at Harvard University, December 1979.

47. International Scientific Forum on Energy for Developed and Developing Countries, Nice, France, October 1979.

48. In many interviews with antinuclear activists carried out for this study, there were requests for anonymity. Fears of reprisals, whether in job loss or revoked passports or working permits, were expressed frequently.

49. D. Nelkin and M. Pollak, *The Atom Besieged: Extra Parliamentary Dissent in France and Germany* (Cambridge, Mass.: MIT Press, 1980).

50. For a more detailed discussion of these issues see: *Nontechnical Issues in Waste Management, Ethical, Institutional, and Political Concerns,* May 1978 (Battelle, Human Affairs Research Centers, Battelle Pacific Northwest Division, Seattle, Washington, 98104; Ref. no. PNL-2400). Also see Ian G. Barbour, *Technology, Environment, and Human Values,* (New York: Praeger, 1980).

51. Brochure from SE$_2$, 570 7th Avenue, Suite 1007, N.Y., N.Y. 10018.

7

When Does Consultation
Become Co-optation?
When Does Information Become
Propaganda?
An Environmental Perspective

Marvin Resnikoff*

From the environmental perspective, there is no simple way to "resolve the social, political and institutional conflicts over the permanent siting of radioactive wastes." On the radioactive waste issue, and more generally, on the nuclear power issue, there is general distrust over the intentions of all federal agencies and government branches and, of course, over the intentions of the nuclear industry. The distrust stems from past and present practices. Yes, we have bad things to say about almost everybody who wants to put a waste dump in our backyard. The "siting issue" will become more of a political issue as time goes on and it will be difficult for true scientists to do disinterested work on the battlefield and for federal agencies to produce factual information which is not propaganda, but this must be done. It will be difficult for federal agencies not to "contaminate" disinterested scientists, those who are conceived by the public to work for the public good, and not to "contaminate" environmental groups, with money, but this must be done. In short, we pose, in this chapter, general concerns that we have and general guidelines for dealing with the public, but pose no technical fix which will "resolve" the conflicts on waste

*With help from Mina Hamilton, Co-project Director, Radioactive Waste Campaign.

disposal. Sorry, but we don't believe that there are any.

Who are "we," presenting the environmental perspective? The Sierra Club is a national environmental and conservation organization. We have 185,000 members in 286 groups in all states of the union. In business since 1892, we are primarily concerned about wilderness protection and Alaska, but we have been active on energy matters and particularly on radioactive wastes. Because of our extensive local structure, we consider ourself a grass-roots organization with a Washington office as well. The Club has been involved, since 1970, with the problem of Nuclear Fuel Services, the now defunct reprocessing plant located in West Valley, New York, but we have also participated in the Pu-recycle (GESMO) proceedings and uranium fuel cycle proceedings before the NRC. In addition to commenting on environmental impact statements, we have underway a major educational, organizing and coalition-building campaign on the radioactive waste issue. This has been proceeding for a year and will be expanding shortly to the northeast, Gulf Coast States, and the Carolinas. The primary goal of the campaign is not so much to reach environmental groups, as middle America. For the Aspen program we have supplied some literature from the campaign. The last edition of our newspaper, *The Waste Paper,* our primary organizing tool, reached 120,000 homes, almost all in New York State. Not everything we do is on paper since we conduct educational/training sessions, lectures, slide shows, newspaper interviews, in every hamlet of the state. We have agreed to participate in this Aspen program to learn better the views of the broad spectrum of persons brought together here and because we thought it would be useful to the folks "back home."

CONFLICT, BUT WHAT ARE THE SIDES?

If there is a conflict, what are the "sides"? What are the issues? The Aspen Institute has brought together the nuclear industry, governors, state legislators, congressmen, federal agencies, lawyers, news media, and the public. We regret that the public is represented here by so few representatives of public interest groups and we feel that this may bias some of the discussions. Each of these groups have a different view, but the sides, when we come down to it, are two: the public and the nuclear industry. And the issue is one, the protection of the health and safety of the public, particularly the public closest to the waste dump. How safe is "safe" depends on where in the spectrum between the public and the nuclear industry one falls. As a generalization, if you are a state legislator, you are much more closely attuned to your constituents than if you are in Washington wrestling with the global problems of the day. If you are an NRC commissioner, you are more closely attuned to the Westinghouse message than the public message because the public does not have scientists who can speak the agency language. (Of course, by Westinghouse, we mean here the nuclear industry). If you are a Supreme Court justice, your latest decisions certainly reflect Westinghouse (recent Price-Anderson and NEPA decisions). If you are President Carter, your last resort has become your first. You even state that nuclear power provides 15 percent of the *energy* of this country, which is what the utilities may think, while it provides only 15 percent of the *electricity* of this country, which is only about 10 percent of the end use energy. In fact, from the local perspective, the three branches of the federal government and associated agencies all resemble the Westinghouse Board of Directors. Even if you are an environmental organization without grass roots and with a global outlook, you are arrayed against local interests in trying to bring high-level waste home. Thus, from our perspective it is Westinghouse and their allies versus local interests and our friends, mainly state legislators and attorneys general. We are skeptical, distrustful. We feel the need to protect ourselves. Given this distrust, the means of protecting ourselves will not be through consensus, which is a very one-sided arrangement, but through the standard means of the democratic process: the vote, the courts, and civil disobedience. Working out how safe is safe in the backrooms of the NRC, DOE, EPA, Battelle, or Keystone is almost impossible given past history, unfolding events and our view of the "sides" in this controversy. The solution is not through consensus bodies, but via the standard democratic processes which will carry out the public will.

HISTORY OF WASTE DISPOSAL—DOES THE NUCLEAR INDUSTRY LEARN FROM PAST MISTAKES?

The nuclear industry, including the weapons industry, has been generating radioactive wastes since 1945, but there is still no solution to the problem of disposing of radioactive wastes. No high-level wastes are being placed in geologic repositories on a commercial basis anywhere in the world. There have been some notable mistakes.

Defense Program

Over 10 percent of the government high level waste tanks have leaked. Some of these tanks at Hanford were older ones. Some of the tanks at Savannah River Plant (SRP) were newer ones of double-walled construction (one of which leaked into the environment). More positively, small amounts of high-level waste from the submarine program have been converted to a solid form, a calcine, since 1963, There have been no leaks, though we have no knowledge of the radiation releases to the environment and the occupational exposure at Idaho. To prevent further leaks at Hanford and SRP, the wastes have been dried to a salt cake within the tank. The salt cake and the liquid wastes will be difficult, if not impossible, to remove from the tanks. The liquid wastes have formed a sludge at the bottom which contains much of the radioactive material. How will the entire system be decommissioned? Will the tanks also go to an underground repository or will they be watched till the radioactivity decays to nonhazardous levels? One would think that the lessons of Idaho are to build a waste calciner, but instead, DOE is buying more tanks for SRP and Hanford.

Perhaps it is worthwhile to add that the federal government has 460 contaminated facilities and sites waiting to be decommissioned and another 100 to become excess in the next five years. Approximately thirty private sites from the Manhattan Project also require further decontamination and decommissioning. There are twenty-two private uranium mill tailings sites that must be "decommissioned," some located within cities. Because of the amount of emitted radioactivity and the proximity of these piles to humans, this is a high priority cleanup problem which the DOE is starting to resolve.

When citizens are informed of the government's handling of its own wastes, they become more questioning of any federal pronouncements and assurances concerning commercial high level waste. Of course, the nuclear industry is quick to point out that the commercial high-level waste management system has improved over its governmental past, but has it?

Lyons, Kansas

This was the AEC's first attempt to bury high-level wastes in salt. However, a neighboring solution mining operation "lost" 175,000 gallons of water. If thermally hot wastes were placed in the Lyons repository, the brine might have migrated to the high-level wastes, perhaps leaching the wastes from the glass or ceramic in a short period of time. It was not the AEC who called a halt to this dangerous experiment, but the public and the state of Kansas, its governor, geological survey, and attorney general. Would the AEC have stopped this project without citizen pressure? What was learned by the AEC in this experience? Did it shed its arrogance and disrespect for the public? As the experience in the state of Michigan suggests, the answer is no.

Carlsbad, New Mexico

The DOE, prompted by the California moratorium, was on a fast track to put commercial high-level wastes in the ground at the Waste Isolation Pilot Project (WIPP) until the 96th Congress restricted funding to an underground TRU "plutonium contaminated" waste respository only. The placement of high-level wastes in salt has come under persuasive scientific criticism—because of its corrosive properties, the fact that water is *attracted* to a heat source in salt and that salt does not effectively absorb radionuclides, its plasticity and its proximity to natural resources, such as oil, natural gas, and potash. Perhaps it should be added that future generations need only *perceive* that natural resources are present for drilling to occur. Unless the borehole were properly plugged, this would provide another migration path to the biosphere. Many scientists and environmentalists believe that a range of geologic media should be investigated before a specific type and specific site is selected. Further, the IRG has recommended a careful, deliberate approach to waste disposal— criteria, regulations, followed by an NRC licensing process. Nevertheless, the WIPP proposal intended to bypass this careful process and was on a fast track. According to DOE's John Deutch, "We hope to submit a license application by 1981, defend it and then submit the license to environmental review and defend that." Fortunately, because of a dispute between congressional committees which oversee defense and DOE spending, a WIPP high-level waste repository was not funded in 1980 fiscal year, and John Deutch did not have his way.

The dispute between these congressional committees is an important one and concerns NRC licensing of these "permanant" waste dumps. The Defense Department does not want NRC licensing because it may jeopardize our national security (it is difficult to take this argument seriously). WIPP was to have served a combined function as a repository for defense TRU wastes from Idaho and 1,000 test commercial spent fuel assemblies. Thus, one level of WIPP would have been licensed by the NRC, the other not. Apparently, it will now only serve for TRU waste. We return to a discussion of these Idaho TRU wastes shortly.

West Valley, New York

West Valley, thirty-five miles south of Buffalo, New York, is the site of the world's first commercial reprocessing facility. Nuclear Fuel Services, a subsidiary of Getty Oil, operated this plant from 1966 to 1972. Because of the high occupational exposures (see the Sierra Club fact sheet, "On the job at NFS") and excessive radiation releases to the environment, the company and the AEC were at continual loggerheads. The public played an important role in publicizing the problems. When the company applied for a construction permit to expand and "modify" their facility, the public intervened. This, plus an increased understanding of the hazards of plutonium separation on a national level, led to a withdrawal of Getty Oil from the reprocessing business in September, 1976. However, Getty Oil has now asked the State of New York to honor contracts signed with the State of New York in 1963 and to assume responsibility for the high-level and low-level wastes remaining on the site. About 60 percent of the wastes came from the defense program and 40 percent from commercial power reactors. The estimated costs to clean up the facility, assuming the technology were available, run to $1 billion. The State of New York, with no technical expertise and financial resources, ran to the federal government for help. The 1980 High Level Waste Solidification Demonstration Project Act will allow DOE to remove and solidify the 560,000 gallons of high level liquid wastes.

From our point of view, the polluters, Getty Oil et al., will be getting off light, but the situation is very serious; the wastes must be removed from the tank as soon as possible. These are highly toxic materials with a very long half-life, sitting in a tank which has a useful life of forty years. DOE and nuclear apologists assure us that corrosion samples indicate that there is no cause for concern. But tanks at Hanford and SRP have leaked at welds due to stress corrosion, or have leaked from the metal plates due to pitting corrosion—neither of which can be predicted by corrosion coupons. It is this type of bland assurance that lacks credibility to the public. Further, a catch pan below the tank has a hole in it eliminating one of the protective barriers.

What has the NRC learned from this experience? Are facilities now designed to aid decommissioning? This certainly was not the case at Three Mile Island. Public interest intervenors have petitioned the NRC to require sufficient funds for decommissioning a facility at any point in its lifetime, but the NRC is resistant to the suggestion. Will hordes of transient employees be used at Barnwell as they were used at NFS? The list is long. The NRC attitude is that NFS was an anomaly, "as low as reasonably achievable" (ALARA) exposure guidelines, and the new dedication of the NRC, as opposed to the AEC, will not allow a repeat performance. But this will occur only if specific, conscious steps are taken by the NRC—a lessons-learned task force, with public input, leading to concrete changes in regulation and design. This was not done. For example, one of the problems at NFS, the need for contact maintenance in "hot" areas of the plant, has been repeated at Barnwell. Instead, the NRC, in the GESMO and uranium fuel cycle proceedings, has presented a model facility where everythings works perfectly, where environmental releases are three to four orders of magnitude less than West Valley experience, if one can believe that, to one part in a billion. The model facility is an "improvement" on the Barnwell facility which has never operated "hot."

Low-Level Waste Burial Grounds at West Valley and Maxey Flats

These burial grounds have leaked radioactivity into the environment. Both states receive high annual rainfalls. Trenches filled with solid low-level wastes: paper, carcasses, contaminated clothes, are cavities which fill up with water and eventually overflow. This occurred at NFS ten years after emplacement of the waste material. A large amount of radioactivity is in both the state-licensed and NRC-licensed burial grounds. For example, 12 pounds of Pu-238 and -239 are in the state-licensed burial ground; 15,000 curies of Sr-90 are located in one trench. How will these burial grounds be decommissioned? Who will provide continual maintenance in perpetuity? Who will pay the costs? What criteria are now available for future low level waste burial grounds? How can the NRC guarantee that these errors will not recur? What lessons have been learned?

This is some of the information that the Radioactive Waste Campaign provides to the public. The federal agencies downplay the past experience, looking toward a brighter tomorrow. As we see it, the nuclear industry and their allies speak of dreams, of idealized facilities, but we in the environmental community like to believe that we are dealing with reality. In this struggle for the hearts and minds of America, what and who does the public believe?

PRESENT PRACTICES
DOE, IRG AND IMPLEMENTATION

Low-Level Waste Dumps

As the Aspen participants are no doubt aware, there is a need for low-level waste burial grounds for medical, research and institutional, industrial, and commercial power reactor radioactive wastes. The IRG report states that the "technologies exist" and that only "existing practice must be improved considerably." The IRG goes on to say that "greater attention to the hydrologic characteristics" must be paid. There surely is no dispute about the "greater attention" phrase, but the IRG statement is not sufficient to bring low-level waste burial grounds on-line, nor to protect the public health and safety. And, of course, the technology of low-level waste burial grounds is very old; it involves digging a hole, throwing waste materials in, and covering the trench with a cover. The technology of digging holes in the ground is quite old. It is also known that cavities in the ground fill up with water, surely more easily than our basements leak. Regions with large rainfalls have leakage problems and erosion.

The solution may be to either locate such burial grounds in a dry region of the country, or in protected above ground storage. It is also important to discuss the need to generate this material. Medical and research use of radioisotopes has been doubling every three to five years. Finally, the long-term management question must be addressed—funding and responsibility for monitoring and maintenance. "The IRG recommends that DOE assume responsibility for developing and coordinating the needed national plan for LLW with active participation . . . from other concerned agencies and *input* from the States, general public and industry in its formulation." This role for the general public, particularly the locally impacted public, of simply providing "input," is entirely unacceptable. Even *local licensing,* rather than NRC or state licensing, does not go far enough.

If DOE were to purchase and manage LLW burial grounds in perpetuity, we believe that the licensing authority should, at the least, remain with the NRC, if not be assumed by local governments themselves. In Germany, the states license nuclear facilities. NRC licensing allows some measure of local protection because the public can intervene in the licensing process. Of course, local citizens should be funded by the federal government to hire scientists and lawyers. In addition, local law and federal law should allow a vote by locals on whether the low-level dump should be located within town and county limits, irrespective of the safety questions. Often NRC proceedings on health and safety issues become forums to decide the desirability of a facility because there is no other forum available. There should properly be a mechanism in federal law which allows a local vote on the simple question of whether a facility is desired, whether it is compatible with the local quality of life.

The long-time scenario for these low-level waste dumps is very clear to us in western New York. Love Canal and the Lake Ontario Ordinance Works site are two familiar examples. LOOW contains uranium mill tailings from the Manhattan Project. Much of the materials are owned by a mysterious company called African Metals, on land owned by the DOE. In both cases, within a twenty-five year period, land which was peripheral to the sites was sold for homes. DOE retains three maintenance personnel on the LOOW grounds, but this will have to take place for 100,000 years. The town of Lewiston would like the site decommissioned so that the land could be put to productive use. The radon levels off-site are greater than maximum permissable concentrations (MPC). Records, after only a twenty-five-year period, are very poorly kept. Many of us have doubts that the concrete tower containing the uranium residues, originally built to hold water, will stand for another twenty-five years because of its inadequate foundation and very heavy load.

In the disaster at Love Canal, homes and schools were built adjacent to the toxic chemical dump. Until barrels began reappearing, no attention was paid to what was buried there nor to a pattern of sickness and cancer that had developed. Now, of course, there is great concern. Law suits in the billions of dollars have been filed against Hooker Chemical, the local school board, and local governments.

We believe that a similar problem will occur at West Valley. It is a rural area at present, but development will come. The State of New York, in this case, will have responsibility for management of the site. The burial ground will have to be maintained and guarded for several hundred years, perhaps longer. Erosion has already taken place and will continue. The club believes that all long-lived materials, particularly the plutonium and the large quantities of Sr-90, should be exhumed so that institutional controls are not required for hundreds to thousands of years. If this is not done, we expect that records will get lost, seepage will occur into future homes, and a new Love Canal situation will arise.

We would be surprised that any local community would accept a low-level radioactive waste dump, or a toxic chemical dump. The IRG may recommend "input," but the real input will be through the democratic process of the vote. We expect that states will side with local interests and governments.

IRG Recommendations

Several of the IRG recommendations and their implementation are of great concern to us. The IRG's call for greater attention to hydrologic characteristics of LLW dump sites is not presently being followed. For example, there is a concerted effort underway to reopen the West Valley site, a site which has already leaked. Surely a site can be found which will not leak. No one in the State of New York has examined alternate sites. The attention to hydrologic characteristics was lacking in 1963 when the West Valley site was opened.

We are concerned that certain sites may not be able to be decommissioned. According to the IRG, "because certain existing sites and/or facilities cannot be decontaminated at a reasonable cost, or perhaps at any cost, long-term institutional control may be required in these exceptional cases." What the IRG is proposing is a double standard for health protection, a different standard for old and new dumps. It tells us that in western New York we are entitled to only a second-rate status in health protection. Long-term institutional responsibility has not worked in the caretaker situations we have seen in western New York; it is simply a method to postpone to some future generation the risks and hard decisions that must be made today.

The IRG's call for public input is being implemented by the DOE by funding environmental organizations to form consensus groups. One such group on LLW dumps is being formed by the Conservation Foundation. It will bring together a small panel of the public and others from different regions of the country. Organizations such as Environmentalists for Full Employment, Natural Resources Defense Council, Solar Lobby, and perhaps others, have also received DOE funds. It would be important to know the full list. We are concerned about the effect of these funds on our friends in the environmental movement and hope that it will not jeopardize their independence.

NRC Regulation of Commercial High-Level Management

The Sierra Club is of the opinion that NRC licensing authority should be extended to the "disposal of waste from both defense and nondefense programs." This is option three of the IRG recommendations. The two types of waste differ in their toxicity and heat production, but the principle we support allow the public control over our environment and local quality of life. It is immaterial the source of the radioactive materials.

In spite of this IRG recommendation, Congress passed legislation which would deny NRC licensing for one aspect of the waste disposal program. Both the House and the Senate passed legislation which would solidify the liquid high-level wastes at West Valley without NRC licensing authority. DOE would simply issue an environmental impact statement, but the public would be locked out of the safety process. No hearings would be required. Further, it is likely that the wastes would be made into a glass before the final repository geologic medium is known. This violates the so-called systems approach to waste management touted by the IRG. Thus, the words of the IRG and its implementation in Congress are diverging from our perspective.

DOE and Engineering Mumbo-Jumbo

The DOE environmental impact statements are almost unreadable by the public, written in an engineering mumbo jumbo that has little physical insight. A certain technocracy has developed with an unbelievable arrogance and disrespect for the public. The flow chart mentality divides a problem into tiny parts without taking a holistic approach. A striking example is the Rasmussen report. The numbers in that document began to assume a reality of their own; engineers believed that absolute probabilities were predicted. But the TMI accident showed that human interaction with machines was not properly modeled, as if it could be.

The modeling of a waste respository is in a much cruder state than the modeling of a reactor and perhaps some engineering humility is in order. It is one thing to know the probability of failure of a SS304 stainless steel pipe, but quite another to know the migration pathways for all the chemical forms in high-level waste, the ion exchange properties of the medium with the contacting water plus leachate. The basic hydrogeologic data on how radioactive material moves through a porous medium is not known. Borehole plugging; how long will it last? Thermal loading; how much should it be? How far will water migrate in salt as a function of the thermal gradient? How many holes can be drilled without undermining the integrity of a repository? The WIPP site has eighty-four holes. Before sites are to be explored, surely this information must be known.

As an example of engineering mumbo jumbo, I have typed a table from the DOE Generic Environmental Impact Statement (GEIS) on the "Management of Commercially Generated Radioactive Waste." In the table, DOE has compared ten commercial waste management concepts with eleven "decision criteria" and twenty-eight "measures of performance." In the last column, I have translated the "attributes" into words that the public understands, into issues that mean something to people. Compare the DOE list with ours. Of course, the DOE also convened a panel of experts to compare these technologies.

We should add that we are also troubled by the IRG recommendations that DOE inform the public "in understandable terms," that is, we are troubled that DOE does not write material understandable to the public and we are worried if they begin to do so because such material may become propaganda

Table 7.1   DOE GEIS "Management of Commercially Generated Radioactive Waste"

Decision Criteria	Attributes	Translation
socioeconomic impact	economic impact on the public sector	big business bailouts by federal government
		corporate irresponsibility in leaving U mill tailings, and waste inflation caused by high capital expenditure on waste disposal
		destruction of future resources by repository
		high federal and state taxes for disposal cost
		high utility rates for disposal cost
	economic impact on the private sector	decrease in land values near disposal site
		tourist revenue decrease near disposal site (Carlsbad, Finger Lakes)
costs of construction and operation	costs of construction and operation	reactor decommissioning cost uncertainties
policy and equity considerations	distribution of risk	deception of public by federal government, utilities
		cost to future generations
		re-allocation of disposal funds to social needs
		military vulnerability
		state rights v. federal govt. civil liberties impacted
	international policy conflicts	importation of foreign wastes
short term radiological safety	anticipated occupational exposure	health effects to workers from handling waste
	accidental non-occupational exposure	hazards of compacted spent fuel pools
		hazardous transportation of wastes
		contamination of water in rivers, lakes and streams
		psychological uncertainty associated with undefined risks
long-term radiological safety	susceptibility to engineer-failures, to natural phenomena, to accidental encroachment	long-term genetic and health effects from leakage

for the nuclear industry as it did during the nuclear referendum in California. The IRG stated,

> The IRG's own experience with public participation and the recommendations of many citizens appearing before the IRG indicate the urgent need for sustained, effective efforts to inform the public. . . .
> Routinely update the status of scientific and technical knowledge on nuclear waste management and provide this information to the public at large in understandable terms.

We support disinterested honest scientific information in the public interest, but we are concerned about efforts to propagandize the public. There must be some public scrutiny of the information that is put out.

DOE Documents Have Become Political Instruments

The DOE EISs have become "politically contaminated science." The scientific arguments are often shaped to fit the wishes of Congress and the policies of the DOE bureaucracy. We cite the following. In proposing that TRU wastes go from Idaho to WIPP, the DOE slants the cost-benefit analysis in that direction by conveniently ignoring a host of costs and discovering benefits that are difficult to take seriously. On the other hand, in proposing that the TRU wastes at West Valley not be removed from the site, the arguments go in the opposite direction. Check this out. In proposing that the Idaho TRU wastes go to WIPP, DOE conveniently ignores the economic costs for transportation and disposal which increase the total costs to $0.75 to $1 billion. This expenditure will save the federal government $0.6 million annually for maintenance of the Idaho TRU wastes "as is" and will protect the public from a possible radiation exposure of 100 person-rems in the event that lava flows over the Idaho site. A lava flow last occurred 400,000 years ago. It is expected that evacuation would mitigate these radiation effects; the evacuation is expected to be rapid in this case. A less convincing argument could not have been written. The real truth lies elsewhere on why the Idaho TRU wastes should be moved. It is in a promise made by AEC Chairman Seaborg to Senator Frank Church in 1970 that the wastes would be moved within a decade.

An opposite picture is presented by DOE in the case of the West Valley TRU wastes; substantial arguments are raised as to why the West Valley TRU wastes should not be moved. In this case, the cost of disposal is $416 million and the cost of transportation is $130 million. The occupational exposures are large; about ten times the person-rems involved in placing the wastes into the LLW burial ground, though several years of decay have taken place. Lava flow is not promised, but it could have been stated that the burial ground did leak radioactivity into the environment within ten years of emplacing wastes in that burial ground, and that the radioactive water does go into Lake Erie and the water supply of Buffalo and environs, with 1.8 million people. In this case,

Congressman Lundine and the nuclear industry would like to reopen the West Valley site for the burial of additional radioactive wastes. DOE provided the EIS which the congressman ordered.

IRG Recommendation on Citizen Participation

As indicated, we are very concerned about the IRG recommendations on citizens participation with the present scientific resources now available to citizens. The process can only work if citizens have some control over the agencies, such as DOE and Battelle, and if citizens can hire their own scientific experts, with federal funds, to oversee the scientific process. The DOE should welcome criticism; citizen groups should hire persons whose job it is to search for problems. There is distrust that government scientists, in an agency that promotes nuclear power, will actually examine the problems critically. The disparity in resources between citizen groups and the nuclear industry, including DOE, is enormous. The government is essentially buying up thousands of scientists about the country to work on waste disposal problems, but environmental groups have only a handful of full-time scientists. One would think that the opposite would be true. After all, Congress passes laws protecting the public health and safety and funding federal agencies to do this job. The federal agencies then evaluate industry applications and hire consultants. The public should feel that there are tens of thousands of scientists working to protect the public health and safety, but the converse is true. Tens of thousands are working for the nuclear industry, the utilities, and their allies in the federal agencies while a handful of scientists work for environmentalists in what we call "the public interest." Citizen groups are almost scientifically defenseless.

An example of the temper of DOE when a scientist bucks the company line is that of Mancuso. This biostatisician was fired when he would not compromise his data for the purposes of DOE. Such disinterested scientists are rare and are highly respected by the public. We are happy that HEW is funding his continuing research. The question that concerns us is, with this enormous amount of money going to DOE, who will protect the grape grower in the Finger Lakes region of New York State?

Consensus Groups

What is the role of consensus groups in the democratic process? While the club has always kept the door open to discussing environmental questions with the nuclear industry and others, inevitably an organization must choose where to place its time and resources, whether it should be in the backrooms of the NRC, DOE, Battelle, Keystone, or directly out with the public. Thus, in general we support the IRG statement, "Support private sector efforts to generate a greater degree of social and technical understanding and agreement on nuclear waste management issues." Some issues which pertain to basic health and safety questions may not be subject to compromise and consensus may not be possible. We are concerned about the tendency of consensus groups to come to agreement and then to margin responsible criticism.

In a democracy, it is not possible to govern without the consent of the governed. There is basic distrust of the federal agencies and whether they will truly protect the health and safety of the public. There is no magic formula to break down this distrust. It will require hard work over several years. It will require honest research by the agencies and the nuclear industry and a frank discussion and not a cover-up of the uncertainties and problems.

Epilogue: Prospects for Consensus

E. William Colglazier, Jr.

The current agenda for nuclear waste management encompasses many significant institutional issues for which a national consensus has not been achieved or implemented. Yet in several areas, progress has occurred toward developing workable national policies and functioning management systems. Ironically, it has been the states, who in trying over several years to capture a role in the decision making, have taken the initiatives that eventually produced these positive results. But much remains to be done, and the history of waste management indicates the formidable challenge of sustaining momentum. This epilogue represents a personal estimate of the prospects for reaching in the near term a stable national consensus in the United States on the unresolved governance issues of nuclear waste management.

LOW LEVEL WASTE DISPOSAL

The passage of the Low Level Radioactive Waste Policy Act in December 1980 established a sound policy framework on which to rebuild an equitable and stable regional management system.[1] The act was the culmination of a year-long effort by several state officials and organizations, including the National Governor's Association and the State Planning Council, to persuade the federal government to let each state be responsible for assuring the safe management of the commercial low-level waste generated within its borders. Through asserting their own self-interest by temporarily closing sites and forcing volume reductions, the three states with operating dumps eventually convinced other states, the generators, and the Congress that a lasting solution to the problem of opening new sites was needed, that it should not come from accepting commercial waste at federal sites, and that states could develop the competence to do the job. The fact that the generators include hospitals and research institutions (generating 25 percent of the volume in 1978), industry (24 percent) as well as commercial power reactors (43 percent), and that these generators maintained a united front in their aggressive lobbying was a contributing factor in persuading other state governments to focus on the issue.[2] The spectre of a possible curtailment of essential and popular services proved to be a powerful incentive to responsible action. But an essential ingredient was the federal government allowing states to generate their own appropriate role and, thereby, to develop a promising solution to a national problem.

The flurry of activity since the passage of the act indicates the seriousness with which some states have accepted their new responsibility. Nevertheless, a sustained effort will be required on the part of the state governments to convert the new policy into practice. The date of January 1, 1986, when a regional compact can exclude waste from non-member states—the key stimulus to action—is not too distant considering the many tasks that need to be completed. By then, the compacts will have to be negotiated, passed by the state legislatures, and approved by Congress. States or compacting regions will have to develop comprehensive management plans and processes for finding new sites. They will have to develop mechanisms for addressing the concerns of local communities and the general public in the site selection process. They will have to choose a competent private operator, establish a financing mechanism, and decide whether the regulator will be the state (under the Agreement States program) or the Nuclear Regulatory Commission.

The strategies for developing new sites may differ by region. Some states may be willing to take the lead in developing a new site and later negotiating a compact with neighbors. In other regions, states may prefer to negotiate a compact before the site selection process is begun. Within several months of the passage of the act, several states in the northwest approved compacting legislation, and a state in the southwest statutorily authorized its own site selection process, for in-state wastes.

Potential hurdles or problems may still lie ahead in the effort to open new low-level waste disposal sites. Finding seed money for starting negotiations and planning during the transition period may be difficult considering the stringent budget situation facing most states. A large state could develop its own site ignoring its neighbors or perhaps be excluded from neighboring compacts, both of which could cause potential difficulties for the lone state or the region. Without adequate contingency planning, states relying solely on a neighbor to develop a site may be left in a difficult situation if the host state cannot deliver. Generators will likely have to consider developing some surge storage capacity in case new sites do not become available when needed. These potential problems are all surmountable if states do not delay in addressing the outstanding issues.

The courts invalidated the Washington state initiative that would have banned incoming shipments and disposal after June, 1981, of nonmedical low-level wastes from states not belonging to a valid compact with Washington.[3]

The presiding federal judge concluded that this initiative violated the supremacy and commerce clauses of the U.S. Constitution. In response, Congress may be asked by the state to decide if the northwest compact can exclude waste from outside its region before 1986, but approval would appear to violate the original congressional intent to leave a reasonable period for making the transition to regional disposal sites.

The federal agencies will continue to have a supporting role to play in low-level waste management. The Department of Energy (DOE) was required by the Low Level Waste Policy Act to prepare a report documenting the status of existing sites and assessing future capacity and transportation requirements. The DOE continues to manage disposal sites for its own low-level waste and perform research and development on new handling and disposal technologies. The federal government may also have to provide technical and financial assistance to states during the transition period. In addition, the Nuclear Regulatory Commission (NRC) will need to finalize its regulations and the waste classification scheme, and the Environmental Protection Agency (EPA) will have to complete its radiation protection standards.

Once a national policy has been accepted by the various levels of government and the public and the relevant technology is reasonably well in hand, the process of implementation should become relatively straightforward. While local communities may need considerable convincing to host a site, the prognosis for the building of a safe regional management system for low-level waste disposal is quite favorable.

HIGH LEVEL WASTE DISPOSAL

The technology for the safe disposal of the exceedingly long-lived and hazardous high-level wastes, which includes reprocessed waste and spent fuel, is complex and untested compared to that for low-level wastes. The importance of stabilizing the governance framework becomes even more critical. As the State Planning Council has emphasized in its final report to the President, the enactment by Congress of comprehensive legislation appears to be essential for firmly establishing and stabilizing national policy for high-level waste management.[4] Although adequate statutory authority exists for assigning the principal role to the federal agencies, executive branch activities have not produced a durable and consistent national policy, and the federal waste management program has suffered from too frequent change in direction. Further change should not be precluded in a program that must span decades, but this must be done through an evolutionary process in order to prevent deterioration of the program's credibility and loss of time.

Comprehensive legislation was nearly achieved in the "lame duck" congressional session in 1980. Due to the results of the election, however, some senators were less interested in accepting the compromises reached in the bill passed the previous summer.They balked at a state consultation and concurrence role for a defense waste repository, particularly as the House conferees would not accept federal away-from-reactor interim storage facilities. Yet, as one House staff aide observed, every side has compromised their principles, so deals can be struck again. It will take time, however, for a new administration to realize that not everything can be controlled through the budget process and federal agency initiatives, and the fragile consensus that existed on many of the issues at the end of the 96th Congress could disappear with further delay.

Among the reasons for enacting new legislation, one of the most important is to secure congressional endorsement of the primary purpose of the federal program. The fundamental question is whether the national strategy should be development of a permanent disposal capability in mined geologic repositories, as DOE is attempting to implement, or reliance on monitored storage in engineered facilities for the foreseeable future, as some congressmen propose. The Department of Energy has completed the final generic environmental impact statement on the management of commercially generated radioactive waste and published a formal record of decision declaring a disposal strategy based upon the development of permanent repositories.[5] Nevertheless, national policies in politically controversial areas are never very secure if based soley on agency directives, even with formal presidential blessings.

The tradeoffs between focusing on permanent disposal or monitored storage as the primary goal of the federal program has been long debated. An engineered facility at or near the surface, designed for retrievable storage for one hundred years or more, is a technically viable option, but one that would require continued institutional monitoring. Quoting an NRC official in comparing the two approaches, "Geologic disposal requires a demanding combination of geologic and hydrologic conditions to assure isolation without human care for many millennia. Long-term surface storage substitutes the institutional requirements of human care and maintenance for the technical requirements of geologic isolation."[6]

Developing criteria to ensure the isolation of nuclear wastes for many thousands of years is surely a demanding technological challenge. Pursuant to its regulatory responsibility, the NRC has issued final procedural criteria and draft technical criteria that are intended to assure public health and safety and to meet the general environmental standards for waste disposal which will eventually be set by the Environmental Protection Agency.[7] (The EPA has been especially tardy by not publicly issuing draft standards to date.) Placing primary programmatic emphasis on the permanent disposal option, however, does not imply making an irreversible decision in the near term. According to the NRC's draft technical criteria, a repository must be designed to guarantee retrievability for approximately one hundred years, so that a decision on closure of a repository will not be made until late in the next century at the earliest. The engineered surface storage facility may be able to guarantee more rapid retrieval, but proceeding now with the development of mined repositories does not require an imprudent or premature foreclosure of options.

Licensing decisions will necessarily be based, in part, on analysis of consequences and events which could potentially disrupt a permanent repository. Consequently, calculations using mathematical modeling will be required to predict performance and give confidence that the wastes will be isolated for the very long time that they will be hazardous. A series of detailed measurements, investigations, and experiments underground during the site characterization process will be required to determine the chemical, mechanical, thermal, and fluid flow properties of the host geological environment in order to compare with earlier lab and field test data[8] and for use as input data for the mathematical modeling associated with performance assessment. Because there will be some uncertainty regarding modeling predictions, the overall philosophical approach for guaranteeing isolation of the wastes is to design multiple barriers, which are to include the waste form, the waste package, the underground facility, and overpack, and the geologic environment. The NRC has proposed setting performance criteria for each barrier, for example, by requiring that the waste package should contain the radionuclides without release for at least the first one thousand years. After that the annual release rate for each radionuclide from the engineered barriers should be no more than one part in one hundred thousand, and the groundwater travel times through the geologic media to the accessible environment should be at least one thousand years. The host geologic environment must also be structurally and tectonically stable, contain only a few surface penetrations, and be unattractive for potential resources (so as to reduce the chance of inadvertent human intrusion). The DOE would prefer to have only an overall performance objective rather than individual objectives for each barrier, but the proposed NRC approach appears to be a reasonable compromise between ensuring multiple barriers and allowing adequate flexibility for implementing a new technology.

Some cooling of the waste will be required for reducing radiation, temperature, and heat load constraints on the waste packaging and the geologic environment. Therefore, some interim surface storage in engineered facilities will always be required, probably for at least ten years after the fuel is taken from a reactor. Some countries may even choose to store the waste for longer periods, perhaps 40 years or more in an interim facility. To many observers, however, it would be unfortunate if renewed interest and funding for a long-term storage option derailed the national focus from development of a permanent disposal capability. The public appears tired of procrastination in developing a permanent solution. The retrievable surface storage facility (RSSF) may be easier to design and site technically, but the siting process may be just as difficult politically. Although the congressional interest in the RSSF does reflect the concerns of some potential host states, it is more difficult to understand why some supporters of nuclear power believe that the long-term storage option can solve their political problems with radioactive waste management.

The second major issue which may be addressed by congressional legislation, although it was avoided in the previously passed House and Senate bills in 1980, is the multiplicity of sites in the characterization process. The Carter administration adopted a technically conservative approach that required the characterization of four to five sites before the first application for a repository construction license could be submitted to the NRC. The purpose was to expand the characterization process beyond salt and beyond the geologic environments found on the federal reservations. The final NRC procedural rules adopted a similar but less restrictive approach by requiring the characterization of at least three sites at repository depth in at least two different rock types in order to support a license application. The DOE held that shafts are not necessary to characterize each medium, so the NRC procedural requirements are significant. The *in-situ* investigations are essential not only for obtaining data to compare with field tests and for use in mathematically modeling, but also for characterizing geological heretogeneities (fractures, faults, breccia pipes, etc.) that might override the average properties of the host rock.

The Reagan administration has adopted a fast-track approach for multiple site characterization by planning to go no further than the minimal NRC requirements. The DOE has accelerated by two years the characterization schedule for bedded salt and volcanic tuff and has focused on three sites for sinking exploratory shafts (basalt on the Hanford Reservation, Washington, tuff on the Nevada Test Site; and a salt site to be selected in 1983 from candidates in Utah, Texas, Louisiana, and Mississippi).[9] It is unlikely that Congress will impose stricter requirements on multiple site characterization than the existing NRC rules.

The third major issue of the high level waste repository development program is the role of an unlicensed test and evaluation (T&E) facility. The Reagan adminstration has proposed that one of the three sites examined at depth would be chosen in 1985 for the construction of a T&E facility.[10] Radioactive waste would be emplaced by 1989, perhaps as much as a decade before the opening of a permanent repository. The characterization process at depth would continue at all three sites, and perhaps others (such as in granite and an additional salt site), as part of DOE's efforts to select one site in around 1988 for application to the Nuclear Regulatory Commission for a construction license for a permanent repository.

The unlicensed T&E facility proposed by DOE is not a small research and development facility, but rather an intermediate sized facility somewhat similar to the intermediate scale repository proposed by the Interagency Review Group (except that the latter facility was intended to be licensed and was not included in the Carter administration's policy statement). The unlicensed T&E facility would be designed for the emplacement and subsequent retrieval of several hundred canisters of solidified high-level wastes in repository depth. The Department of Energy proposes that such a facility need not be licensed since it would be a temporary facility intended for developing early engineering experience in repository construction and large waste handling operations. The DOE would attempt to preserve the option that the T&E site could be considered later for licensing as a permanent repository.

The conventional NRC approach is to characterize a minimum of three sites at repository depth, as required by the procedural regulations, without building a T&E facility. The number of thermally "hot" canisters, either heaters or actual radioactive waste, needed for "in-site" characterization measurements would probably be no larger than the amounts used in prior field tests, which is considerably less than the number envisioned for emplacement in a T&E facility. After DOE receives a construction license for one of the qualified sites, the first development phase of the permanent repository would then be very similar to the activities undertaken at a T&E facility.

The T&E facility proposed by DOE would emplace a fairly large amount of high level waste deep underground perhaps a decade before a permanent repository is open, which may have some symbolic value. Because the information generated by a T&E facility is not required for a site to receive a construction license, however, it is not necessary for meeting the NRC's criteria for making predictions about a repository's performance far into the future. Because of the necessity to use mathematical modeling to project long-term repository performance, the significance of a T&E facility for waste disposal is quite different from a conventional demonstration project which occurs over a timeframe that is similar to the real facility's lifetime performance. The T&E project could be useful for improving the design of a future permanent repository. It could also be used to work out some detailed licensing issues without having to go through the formal licensing process. The cost, however, would be relatively high, perhaps on the order of $300 to $500 million as opposed to $30 to $50 million for sinking a shaft and characterizing a site at depth.

The potential disadvantages of a T&E facility compared with the conventional approach include having to make large expenditures for project construction which could reduce funding for other parts of the DOE program in this decade, such as for multiple site characterization. The site might also be compromised by unalterable changes during T&E project construction which could then prevent its consideration for a permanent repository. This eventuality would be especially likely if design and construction proceed on a fast-track schedule that inhibits proper preparatory work. If the T&E project does not become the site of a permanent repository, then a similar and perhaps even duplicate expenditure would be required during the early development phase of the first permanent repository. The greatest liability of the T&E project could be exacerbation of fears by the states and the public that the federal government is attempting to bypass institutional safeguards for siting a permanent repository. In other words, there will be concern that the expense and bureaucratic momentum tied up in a T&E facility may result in a future Congress and Administration forcing that facility to be the first permanent repository. However, if the Congress believes that this symbol of emplacing large amounts of waste in the ground in this decade is important enough to justify the expense of a T&E facility, and is willing to define sufficient institutional safeguards to assure that a T&E facility would not become a de facto permanent repository, then the T&E approach is not necessarily inconsistent with a technically conservative repository development program.

The fourth and probably the most important issue for new legislation to resolve is the role of states and Indian tribes with respect to the repository development program. The State Planning Council has proposed a strategy to guide the federal agencies and potential host states and tribes through the repository siting process.[11] The strategy incorporates an intergovernmental partnership, which has been labeled consultation and concurrence, for state and tribal participation. Consultation and concurrence with the implementing agencies would involve the negotiation of written agreements, the timely sharing of information, independent technical review by the state, and a process of incremental decision making designed to build consensus on site specific issues.

Recognizing that impasses may occur between the federal government and an individual state or tribe, the Council proposed a federal mechanism to resolve these conflicts to guarantee that decisions can be made on the siting of a repository. The host state or tribe would not be given an absolute veto over the siting process, but rather it would possess the right to have its objections at two key decision points considered formally by the president or the Congress. A state or tribe would be allowed to petition the president to review its objections to the selection of a site for sinking a shaft and characterizing the geologic environment at depth. Similarly, they would be permitted to petition the Congress for a formal review of their objections to a decision by DOE that a site is qualified for a repository. The Council recommended that the state or tribal objections be sustained at this point unless overridden by both houses of Congress.

Once ratified in legislation, the Council's detailed partnership approach could produce timely answers and settle disputes and still meet the legitimate concerns and needs of all levels of government. It could also minimize the threats of arbitrary exercises of political power by one governmental entity against another and perhaps avoid having to resort to judicial resolution.

The Council also recommended that the consultation and concurrence process be applied to the siting of a test and evaluation facility, which the

Council believed should only be constructed if Congress determines that such a facility is technically justified in support of the overall program for the development of a permanent repository. The Council proposed that any statutory authorization for a T&E facility include an upper limit on the amount of waste temporarily emplaced and a requirement that after a specified period of time the wastes be withdrawn unless the facility has been licensed by the NRC.

The bills that separately passed the House and Senate in the 96th Congress included provisions for a congressional override of a state's objections to the siting of a repository for commercial high level wastes. These bills required that a state's objections be upheld only if one house of Congress affirmatively concurred by passing a special resolution. This mechanism, which is the controversial "one house veto" of a presidential directive (presuming that the president supports the siting decision of his agency), places the burden on the state for obtaining congressional action to sustain its point of view. From the state perspective, this override mechanism is slightly weaker than the provision endorsed by the State Planning Council. The bill that passed the Senate in the summer of 1980 distinguished in the consultation and concurrence process for a defense repository by requiring in the latter case affirmative action by both houses on Congress to uphold a state's objections, which is called "the two house veto" of a presidential decision. The nuclear industry has preferred this requirement even for commercial repositories, which is the strongest provision from the executive branch perspective.[12]

In all these provisions, the state objections would come prior to the consideration by the NRC of a license application. Some federal officials favor congressional consideration of state objections after all the technical information is available, that is, after the licensing process is completed. The State Planning Council felt that this decision point was so late in the process, with the added bureaucratic momentum and expended funds, that it would be unlikely for Congress to give serious consideration to a state's objections. Significant technical information would be available following the characterization of a site at depth, which is the point in time that the State Planning Council seeks statutory authority for a state to submit its formal objections to the Congress.

Whatever congressional override mechanism is adopted, the importance of enacting legislation is to formalize the consultation and concurrence process. Then the long road of implementation of an intergovernmental partnership can proceed in a constructive and cooperative atmosphere where all parties are secure in their rights and aware of the rules. The Reagan administration has not been generally favorable to the need for new legislation in high-level waste management, but DOE officials have indicated that they may seek legislative guidance to provide for formal resolution of possible state objections to the siting of a repository.

Even if new congressional legislation formalizes the consultation and concurrence process, further actions are required by the executive branch agencies to assure its implementation. In particular, the Department of Energy will have to complete in good faith negotiations with all potential host states and tribes in order to have in place written agreements outlining the details of the consultataion and concurrence process. The NRC would not be able to share with states its statutorily-defined regulatory responsibility for high-level waste disposal. which includes setting criteria and standards and ruling on license applications. Similar restrictions would apply to the Department of Transportation and the Environmental Protection Agency concerning their regulatory responsibilities. However, the regulatory agencies must listen carefully to state and public concerns and provide clear guidance for their role in the various licensing and rule-making processes.

States and tribes must decide how and when to organize themselves for interaction with the federal agencies. For potential host states, this will require setting up a policy task force or review panel to represent the state or tribe in negotiations and consultations with the agencies. States and tribes also need to provide themselves with adequate technical capability. State objections to agency decisions will likely be seriously considered by Congress only if they are for legitimate and substantive grounds. The true test of the intergovernmental partnership approach will occur as it is exercised under actual political and economic conditions at a particular site.

The existence of the State Planning Council, which lasted for eighteen months and expired in August 1981, provided a symbol of federal willingness to listen to state concerns. It also demonstrated the utility of having some institutional mechanism for building a cooperative and non-adversarial approach between the states and the federal government. The states were given a forum for defining a responsible role for themselves without the burden of one-on-one encounters with federal agencies on site specific issues. The existence of the council also aided the federal agencies. It provided a vehicle for soliciting state support for federal programs and policies and a focal point for resolving broad state concerns. The opportunity to lower the political temperature in potential host states and to defer or avoid court action can only enhance the progress of the waste disposal program. The test of a lasting contribution from the council awaits legislative action on its conceptual guidelines and conflict resolution mechanisms for guiding state and federal interactions.

Although the Carter administration asked that the council be authorized in legislation for an extended lifetime, the council did not seek to perpetuate itself. It did recognize the need for continuing some formal dialogue or facilitating mechanisms between states, tribes and the federal government concerning radioactive waste management. The Reagan administration has not yet addressed whether any formal intergovernmental councils to further the work of the State Planning Council are needed.

The next major issue that needs to be resolved in congressional legislation is whether to subject repositories for disposal of defense wastes to regulatory or other requirements which are the same as those for permanent disposal of similar civilian wastes. Currently, NRC licensing is required for facilities for long-term storage or permanent disposal of commercial and defense high-level waste and commercial transuranic (TRU) waste but not defense transuranic waste. According to the draft technical criteria of the NRC, the long-term performance requirements for containment of TRU wastes in a permanent repository are the same as those for high-level waste. From the perspective of many states and their citizens, the institutional safeguards in the repository development program should be applied uniformly to all radioactive wastes of equivalent hazard. Therefore, the State Planning Council and the Interagency Review Group supported the extension of NRC licensing authority to all repositories for the permanent disposal of transuranic waste, including the Waste Isolation Pilot Plant in New Mexico. Similarly, the NRC was encouraged by the council to consult extensively with the DOE over any unlicensed research or test and evaluation facilities designed to test the emplacement of defense high-level or TRU waste underground.

The State Planning Council also supported the application of its consultation and concurrence process to all repositories regardless of the institutional origin of the wastes. The small, but significant difference in the appeal process between "consultation and concurrence" and "consultation and cooperation" (the Congress required only the latter for the WIPP facility) is that the host state can raise the key siting issues to the president and the Congress for the former process and to the secretary of energy for the latter. A limited extension of the NRC licensing authority and a consultation and concurrence process to the DOE repository program for defense wastes should not compromise in any way the national security interests of the United States.

Nevertheless, the extension of NRC licensing and consultation and concurrence to repositories for defense wastes is a very controversial issue in Congress. Through the action of the Armed Services Committees, Congress ruled that the WIPP facility would not be licensed by the NRC, would not receive commercial waste, and would not be subject to a concurrence right for New Mexico. These committees were even able to force the separation of the defense and commercial repository programs within the DOE, which is an additional budgetary expense and management inconvenience. Although New Mexico and the DOE were finally able to sign a consultation and cooperation agreement and temporarily settle the suit brought by the state Attorney General in 1981, the current effort by the DOE in vigorously pursuing WIPP could exacerbate relations between the federal government and potential host states and could result in a repeat of the Lyons, Kansas experience. Other potential host states will be looking at the WIPP experience as a forerunner of how they will be dealt with.

If necessary for resolving this contentious issue, the compromise reached in the earlier Senate bill, which would have authorized a slightly different impasse resolution mechanism between a state and the federal government on a defense repository (requiring both houses of Congress to support the state's objection for the veto to be sustained), would probably be acceptable to states. If resolution of the defense issue is not possible, the 97th Congress could still define statutorily the consultation and concurrence process for commercial waste disposal and defer the defense repository question to a later Congress. The Reagan administration has stated that any new legislation should be directed only to the disposal of civilian wastes and that defense program activities should not be subjected to control by an independent regulatory agency.[13]

Other issues may be addressed if new legislation on high level waste management emerges from Congress. One important issue is the mechanism for financing and allocating costs for the development of a permanent repository. Although there is a difference of opinion on details, the basic philosophical approach agreed to by most is that repository costs should be paid by the generators and borne by the beneficiaries of the waste producing activities, whether it be for electricity or national security. The federal government will obviously pay for disposal of defense wastes and probably for most of the research and development. Many environmentalists and economists prefer a tax on nuclear generated electricity to provide a fund to pay the commercial sector's share of repository development. This mechanism would insure that current beneficiaries begin paying the costs now. However, it is difficult to estimate the cost of permanent disposal ahead of time, and energy taxes are not politically popular. Therefore, it may not be practical to have any option other than the federal government bearing repository costs initially and then recovering costs for commercial waste disposal from the generators when the waste is transferred to the permanent repository. This is the option supported by the industry.

A mechanism is also needed to authorize the negotiations with states for federal compensation of socioeconomic impacts associated with repository development. This latter issue will likely be crucial to gaining acceptance from a host state once safety concerns are allayed.

Many congressional bills, the industry, some environmentalists, and the State Planning Council appear to have a fixation on statutorily defined schedules for the repository development program. The industry seeks to force progress, the states to insure that temporarily stored wastes are eventually moved, and some environmentalists to link failure in meeting schedules with the shutdown of nuclear reactors. The Department of Energy has always opposed mandatory schedules, and the delays by the EPA in issuing standards indicates that agencies can always claim unforeseen events or the lack of adequate staffing or funding for not meeting schedules. Nevertheless, it is likely that target schedules for completing key elements of the federal program will appear in a bill that emerges from Congress.

Another key issue is the improvement of coordination among the federal agencies. The lack of interagency coordination has been a continuing difficulty and is seen by outsiders as a key institutional weakness of the federal program. Several groups have recommended that the overall waste management effort be aggressively directed by a high-level interagency management committee, including participation by the White House and the Office of Management and Budget.[14] This committee might be more effective in extracting and enforcing real commitments from the agencies on improved coordination. An interagency working group on waste management has been created, but its membership is not at the political appointment level. Except for the agreement between the NRC and DOT, the agencies have not yet issued in writing the various bilateral memoranda of understanding outlining their working relationships with each other.

The organization of the federal waste management effort could also change if the Department of Energy should be dismantled, as proposed by President Reagan. The executive branch and the Congress would then have to carefully consider alternative organizational options with special emphasis on preserving the program's continuity, purpose, and accountability.

Several congressional bills have also addressed the preparation of a national plan on radioactive waste management, a process which was undertaken by the Carter administration. The process of preparing a national plan could be a useful vehicle for improving coordination among the federal agencies, incorporating input from states and tribes, and eliciting public participation. As the State Planning Council had recommended, the Carter administration circulated a draft of the national plan to state governments and Congress for their input in order to produce a revised draft for formal public review. The Reagan administration received these comments from the states, but has not yet decided whether to proceed with the development of a national plan document. The importance of integrated long-range planning might be sufficient reason not only for the new administration to proceed with the plan, but also for the numerous congressional committees with relevant jurisdiction to at least consider developing a more coordinated manner for reviewing the federal program, such as by holding joint hearings or forming an ad-hoc joint oversight committee in each house.

The State Planning Council concluded that public participation in waste management planning and programs is sufficiently important to deserve the same quality of thought, commitment, and implementation as technical programs. The formal processes for public participation now include commenting on environmental impact statements and participating in some licensing hearings. Obviously, there are many informal ways that certain members of the public actively participate in and significantly affect the conduct of the federal programs. The Reagan administration does not appear receptive to additional formal mechanisms for public participation.

Legislation on the management and disposal of high-level waste was nearly consummated in the last Congress. Whether the outstanding issues will be resolved in legislation by the 97th Congress remains to be seen, but it is clear that an urgent need exists to develop a statutory framework for firmly establishing national policy.

INTERIM STORAGE AND REPROCESSING

The reference nuclear fuel cycle envisioned ''by the pioneers in the nuclear community decades ago'' incorporated interim storage of spent fuel, reprocessing spent fuel to recycle the unused uranium and plutonium, immobilization of the fission products, temporary storage of the solidified wastes, and then disposal in a geologic repository.[15] This closure of the nuclear fuel cycle was put in abeyance by President Ford's announcement in October, 1976 that "the United States should no longer regard reprocessing of used nucléar fuel to produce plutonium as a necessary and inevitable step in the nuclear fuel cycle, and that we should pursue reprocessing and recycling in the future to be consistent with our international non-proliferation objectives." In April, 1977 President Carter said, "we will defer indefinitely the commercial reprocessing and recycling of the plutonium produced in U.S. nuclear power reactors." Because spent fuel was expected to be filling up utility storage pools in the early 1980s, and society could not allow operating reactors to shut down for lack of storage space, the Carter administration promised to seek congressional authorization for taking title to utility spent fuel that would be delivered to federal away-from-reactor (AFR) storage facilities.

Congress never took action on federal AFRs, although a provision was included in the Senate bill passed in 1980. Although the Reagan administration expressed its support for commercial reprocessing, it disapproved a request from the Secretary of Energy to buy and operate the mothballed Barnwell facility either for a demonstration of commercial reprocessing or the first installment of federal AFR storage capacity. Subsequently, DOE announced that it had discontinued its efforts to provide federal AFR storage facilities.[16] The date when new storage capacity is projected to be first needed has been pushed back to 1986 as nearly all utilities are planning to re-rack existing pools to provide for denser storage. Several years additional lead time to develop new capacity would become available if intrautility or interutility transshipment between existing storage basins and temporary relaxation of full core reserve are permitted.

The State Planning Council agreed with the policy of the Reagan Adminsitration in recommending that utilities should continue to be responsible for the interim storage of their spent fuel. Although utilities still seek to have federal AFRs, the most important factor in utilities being able to meet their interim storage responsibility is maintenance of a constant policy by the federal government. The utilities would then have the certainty and incentive to provide for their own capacity. Either at-reactor or away-from-reactor storage is technically viable and licensable, but storage at reactor sites is probably easier to implement politically as the Tennessee Valley Authority has concluded.[17] Thus, the initial storage of spent fuel is an area in waste management where new legislation is not necessary to convert the status quo into a national policy consensus.

The federal government will need to work cooperatively with industry to develop and demonstrate new storage technologies, such as dry storage or rod storage, which may offer superior and cheaper alternatives to existing pool technology. The NRC will need to observe closely these efforts to facilitate eventual licensing of acceptable new technologies. All the federal agencies must proceed expeditiously with the programs for the development of a permanent repository, so that the utilties' responsibility for interim storage is not open ended.

To guarantee an uninterupted supply of electricity to their citizens at reasonable cost, state governments were encouraged by the State Planning Council to take reasonable steps to facilitate utility efforts to develop or share storage capacity. Utilities may encounter institutional problems which can be alleviated by actions of state authorities, either acting formally or acting as an intermediary.

The new administration's commitment to reprocessing is indicated by its statement that "the cornerstone of the waste management program should be that the reference waste form, as it was prior to the Carter administration and as is in concert with the rest of the world, is reprocessed high-level waste . . . national reprocessing capability is the key to the formulation of our high level waste programs."[18] Although the Reagan administration desires to return to the halcyon days of the reference nuclear fuel cycle, it is not clear where that commercial reprocessing capability is going to come from. Reprocessing is not commercially viable at present because the private sector is unwilling to accept the financial risk and the regulatory uncertainty of commercial reprocessing ventures on its own. Neither the Reagan administration nor many state officials favor the industry's proposal of federal subsidization or demonstration of commecial reprocessing. Governor Riley has stated that in his opinion the biggest environmental problem in South Carolina is the large quantities of temporarily-stored liquid high-level waste from defense activities at the Savannah River facility, which came from reprocessing. He is not in favor of commercial reprocessing until permanent disposal of existing high-level waste is available.

The International Nuclear Fuel Cycle concluded that "the relative differences of the impacts from waste management and disposal between (different) fuel cycles are not decisive factors in choosing a fuel cycle."[19] In other words, permanent repositories can be designed to dispose of both spent fuel and solidified high-level waste. Nevertheless, the Department of Energy has proposed the emplacement of only solidified high-level waste in its testing and evaluation facilities. The nuclear industry would prudently prefer that demonstration facilities include spent fuel emplacement as well. Because commercial reprocessing appears far off for all practical purposes, it would appear to be shortsighted for the new administration to ignore development of a dual disposal capability for spent fuel and solidified high-level waste in permanent repositories.

TRANSPORTATION AND OTHER ISSUES

The transportation of radioactive wastes is another area where new legislation is not necessary to achieve a national policy consensus. All that appears to be required to guarantee a stable and permanent framework is refinement of existing policy. An appropriate level of national uniformity in routing radioactive materials, generally utilizing the interstate highways as preferred routes, has been established by the rule-making procedure initiated by the Department of Transportation (DOT). The DOT rule on highway routing will become effective on February 1, 1982, and will permit preemption of conflicting state and local ordinances.[20] As the price for acquiescing to a uniform national system, state and tribal governments have requested that they be allowed to receive advance notification of high-level or large quantity shipments and be assisted in developing processes for designating alternate routing where needed.[21] The timely implementation of these two reasonable requests would accelerate the ongoing process of developing a national transportation management system that can protect public health and safety and preserve the free flow of commerce.

The Department of Transportation deferred consideration of promulgating a national prenotification rule pending the outcome of the NRC rule-making on the same topic, which is in the process of being finalized.[22] An appendix of the DOT routing rule, however, requires existing state or local prenotification rules to be held inconsistent. The State Planning Council opposed federal preemption of local requirements prior to the implementation of a national uniform system for prenotification.

The DOT rule also authorized processes for considering alternate preferred routes when a state or local community believes that there exists a more acceptable and safer alternative. The designation of an alternate route requires a process of consultation with the affected local jurisdiction and a justification that the new route minimizes overall risk to the public. The DOT will need to develop guidelines and provide assistance to states and localities in developing the processes for selecting alternate routes.

Other areas in nuclear waste management that may need to be addressed in the near term by United States governance institutions include: the clean up of inactive mill tailings, the decommissioning and decontaminiation of retired nuclear facilities, the sub-seabed disposal of high-level wastes, the ocean-dumping of low-level wastes, and the acceptance of foreign spent fuel for nonproliferation purposes. The International Nuclear Fuel Cycle Evaluation concluded that the environmental and radiological impacts of the nuclear fuel cycle are mainly correlated to the mill tailings.[23] The passage of the Uranium Mill Tailings Radiation Control Act of 1978 authorized DOE to enter into cooperative agreements to clean up uranium mill tailings at certain inactive sites. But, as the industry claims in a suit over the regulations for cleaning up active sites, the costs may be quite large for rather uncertain benefits. Concerning decommissioning, Governor Richard Thornburgh of Pennsylvania has stated that the clean up at the disabled Three Mile Island reactor and removal of the nuclear wastes is "the acid test for nuclear power in the United States."[24] For ocean disposal, many islands in the pacific basin are opposed to Japanese plans for ocean dumping of low-level wastes. Opposition will, also, surely grow as countries lacking suitable geological sites for land repository seriously consider sub-seabed disposal of high-level wastes. The Israeli attack on the Iraqi research reactor in 1981 was a watershed event, like the Indian nuclear explosion in 1974, in focusing world attention on the dangers and instabilities caused by the threat, and the responses to the threat of countries possibly diverting enriched uranium or plutonium from peaceful nuclear facilities to make nuclear weapons. These last two issues may continue to generate international political problems at least as contentious as the domestic political problems with nuclear waste disposal on land.

CONCLUSION

This personal assessment of the outstanding governance issues and the prospects for reaching a national consensus may appear optimistic considering the troubled history of nuclear waste management. Yet significant progress over the last few years has occurred in the United States through the sustained effort and responsible actions taken by various government officials and private interest groups. The period of reassessment that occurs with every new administration could result in radical change that voids emerging consensus and again sets back the program. If, however, we rely on our participatory democracy and the progress made to date as a guide, our governance institutions may soon complete the stable organizational framework needed for constructing a safe and societally acceptable solution to the radioactive waste problem.

NOTES

1. Public Law 96-573.

2. Information on low-level waste is contained in: National Low-Level Waste Management Program, *Managing Low Level Wastes: Proposed Approach,* LLWMP-1, August 1980, and *Understanding Low-Level Radioactive Waste,* LLWMP-2, November 1980 (EG&G Idaho, Inc., PO Box 1625, Idaho Falls, ID 83401).

3. "State's Nuclear Waste Ban is Ruled Unconstitutional," *New York Times,* June 27, 1981.

4. State Planning Council on Radioactive Waste Management, *Report to the President,* August 1981.

5. Federal Register 26677, May 14, 1981.

6. William Dirks, Director of Operations of the Nuclear Regulatory Commission, testimony before the Subcommittee on Energy and the Environment of the House Interior and Insular Affairs Committee, June 25, 1981.

7. The NRC's final procedural criteria are published in Federal Register 37: 13971-13987 (February 25, 1980).

8. Field tests have been conducted at: Project Salt Vault (Kansas), Avery Island (Louisiana), and the Asse Mine (Germany) in salt; the Climax (Nevada) and Strippa (Sweden) mines in granite; and the Near Surface Test Facility (Washington) in basalt.

9. Kenneth Davis, Deputy Secretary of Energy, testimony before the Subcommittee on Energy and the Environment of the House Committee on Interior and Insular Affairs, July 9, 1981.

10. Ibid.

11. State Planning Council, 1981.

12. Sherwood H. Smith, Jr., testimony on behalf of the American Nuclear Energy Council, the Edison Electric Institute, and the Utility Waste Management Group to the Subcommittee on Energy and the Environment of the House Committee on Interior and Insular Affairs, July 9, 1981.

13. Davis, testimony before House Committee. See also the White House policy statement, Oct. 8, 1981.

14. Task Force on the National Plan, Report to the State Planning Council, sent to the President on January 13, 1981 and published in the appendix to the Council's *Report to the President,* op. cit.

15. Davis, testimony before House Committee.

16. Letter from Omer Brown, Office of the General Counsel, Department of Energy, to Marshall E. Miller, Administrative Judge, U.S. Nuclear Regulatory Commission, March 27, 1981.

17. Hugh G. Parris, Manager of Power, Tennessee Valley Authority, testimony to the subcommittee on Energy and the Environment of the House committee on Interior and Insular Affairs, June 25, 1981.

18. Davis, testimony before House Committee.

19. International Nuclear Fuel Cycle Evaluation, *Waste Management and Disposal,* Report of Working Group Seven, INFCE/PC/2/7, February 25, 1980, p. 13.

20. Federal Register 12: 5298-5318 (January 19, 1981). For a survey of the issues, see Robert W. Bishop, "Transportation of Radioactive Materials: A Lawyer's View of State and Federal Regulations," presented to the American Nuclear Society.

21. State Planning Council, 1981.

22. Federal Register 46: 13971-13987 (February 25, 1981).

23. International Nuclear Fuel Cycle Evaluation, op. cit.

24. "U.S. Aid Sought for Reactor Cleanup." *New York Times,* July 10, 1981.

Appendix A: Presidential Message and Fact Sheet of February 12, 1980

OFFICE OF THE WHITE HOUSE PRESS SECRETARY

THE WHITE HOUSE

TO THE CONGRESS OF THE UNITED STATES:

Today I am establishing this Nation's first comprehensive radioactive waste management program. My paramount objective in managing nuclear wastes is to protect the health and safety of all Americans, both now and in the future. I share this responsibility with elected officials at all levels of our government. Our citizens have a deep concern that the beneficial uses of nuclear technology, including the generation of electricity, not be allowed to imperil public health or safety now or in the future.

For more than 30 years, radioactive wastes have been generated by programs for national defense, by the commercial nuclear power program, and by a variety of medical, industrial and research activities. Yet past governmental efforts to manage radioactive wastes have not been technically adequate. Moreover, they have failed to involve successfully the States, local governments, and the public in policy or program decisions. My actions today lay the foundation for both a technically superior program and a full cooperative Federal-State partnership to ensure public confidence in a waste management program.

My program is consistent with the broad consensus that has evolved from the efforts of the Interagency Review Group on Radioactive Waste Management (IRG) which I established. The IRG findings and analysis were comprehensive, thorough and widely reviewed by public, industry and citizen groups, State and local governments, and members of the Congress. Evaluations of the scientific and technical analyses were obtained through a broad and rigorous peer review by the scientific community. The final recommendations benefited from and reflect this input.

My objective is to establish a comprehensive program for the management of *all* types of radioactive wastes. My policies and programs establish mechanisms to ensure that elected officials and the public fully participate in waste decisions, and direct Federal departments and agencies to implement a waste management strategy which is safe, technically sound, conservative, and open to continuous public review. This approach will help ensure that we will reach our objective—the safe storage and disposal of all forms of nuclear waste.

Our primary objective is to isolate existing and future radioactive waste from military and civilian activities from the biosphere and pose no significant threat to public health and safety. The responsibility for resolving military and civilian waste management problems shall not be deferred to future generations. The technical program must meet all relevant radiological protection criteria as well as all other applicable regulatory requirements. This effort must proceed regardless of future developments within the nuclear industry—its future size, and resolution of specific fuel cycle and reactor design issues. The specific steps outlined below are each aimed at accomplishing this overall objective.

First, my Administration is committed to providing an effective role for State and local governments in the development and implementation of our nuclear waste management program. I am therefore taking the following actions:

- By Executive Order, I am establishing a State Planning Council which will strengthen our intergovernmental relationships and help fulfill our joint responsibility to protect public health and safety in radioactive waste matters. I have asked Governor Riley of South Carolina to serve as Chairman of the Council. The Council will have a total of 19 members: 15 who are Governors or other elected officials, and 4 from the Executive departments and agencies. It will advise the Executive Branch and work with the Congress to address radioactive waste management issues, such as planning and siting, construction, and operation of facilities. I will submit legislation during this session to make the Council permanent.

- In the past, States have not played an adequate part in the waste management planning process—for example, in the evaluation and location of potential waste disposal sites. The States need better access to information and expanded oportunity to guide waste management planning. Our relationship with the States will be based on the principle of consultation and concurrence in the siting of high level waste repositories. Under the framework of consultation and concurrence, a host State will have a continuing role in Federal decisionmaking on the siting, design and construction of a high level waste repository. State consultation and concurrence, however, will lead to an acceptable solution to our waste disposal problem only if all the States participate as partners in the program I am putting forth. The safe disposal of radioactive waste, defense and commercial, is a national, not just a Federal, responsibility.

- I am directing the Secretary of Energy to provide financial and technical assistance to States and other jurisdictions to facilitate the full participation of State and local government in review and licensing proceedings.

Second, for disposal of high level radioactive waste, I am adopting an interim planning strategy focused on the use of mined geologic repositories capable of accepting both waste from reprocessing and unreprocessed commercial spent fuel. An interim strategy is needed since final decisions on many steps which need to be taken should be preceded by a full environmental review under the National Environmental Policy Act. In its search for suitable sites for high level waste repositories, the Department of Energy has mounted an expanded and diversified program of geologic investigations that recognizes the importance of the interaction among geologic setting, repository host rock, waste form and other engineered barriers on a site-specific basis. Immediate attention will focus on research and development, and on locating and characterizing a number of potential repository sites in a variety of different geologic environments with diverse rock types. When four to five sites have been evaluated and found potentially suitable, one or more will be selected for further development as a licensed full-scale repository.

It is important to stress the following two points: First, because the suitability of a geologic disposal site can be verified only through detailed and time-consuming site specific evaluations, actual sites and their geologic environments *must* be carefully examined. Second, the development of a repository will proceed in a careful step-by-step manner. Experience and information gained at each phase will be reviewed and evaluated to determine if there is sufficient knowledge to proceed with the next stage of development. We should be ready to select the site for the first full-scale repository by about 1985 and have it operational by the mid-1990's. For reasons of economy, the first and subsequent repositories should accept both defense and commercial wastes.

Consistent with my decision to expand and diversify the Department of Energy's program of geologic investigation before selecting a specific site for repository development, I have decided that the Waste Isolation Pilot Plant project should be cancelled. This project is currently authorized for the unlicensed disposal of transuranic waste from our National defense program, and for research and development using high level defense waste. This project is inconsistent with my policy that all repositories for highly radioactive waste be licensed, and that they accept both defense and commercial wastes.

The site near Carlsbad, New Mexico, which was being considered for this project, will continue to be evaluated along with other sites in other parts of the country. If qualified, it will be reserved as one of several candidate sites for possible use as a licensed repository for defense and commercial high level wastes. My fiscal year 1981 budget contains funds in the commercial nuclear waste program for protection and continued investigation of the Carlsbad site. Finally, it is important that we take the time to compare the New Mexico site with other sites now under evaluation for the first waste repository.

Over the next five years, the Department of Energy will carry out an aggressive program of scientific and technical investigations to support waste solidification, packaging and repository design and construction including several experimental, retrievable emplacements in test facilities. This supporting research and development program will call upon the knowledge and experience of the Nation's very best people in science, engineering and other fields of learning and will include participation of universities, industry, and the government departments, agencies, and national laboratories.

Third, during the interim period before a disposal facility is available, waste must and will continue to be cared for safely. Management of defense waste is a Federal responsibility; the Department of Energy will ensure close and meticulous control over defense waste facilities which are vital to our national security. I am committed to maintaining safe interim storage of these wastes as long as necessary and to making adequate funding available for that purpose. We will also proceed with research and development at the various defense sites that will lead the processing, packaging, and ultimate transfer to a permanent respository of the high level and transuranic wastes from defense programs.

In contrast, storage of commercial spent fuel is primarily a responsibility of the utilities. I want to stress that interim spent fuel storage capacity is *not* an alternative to permanent disposal. However, adequate storage is necessary until repositories are available. I urge the utility industry to continue to take all actions necessary to store spent fuel in a manner that will protect the public and ensure efficient and safe operation of power reactors. However, a limited amount of government storage capacity would provide flexibility to our national waste disposal program and an alternative for those utilities which are unable to expand their storage capabilities.

I reiterate the need for early enactment of my proposed spent nuclear fuel legislation. This proposal would authorize the Department of Energy to: (1) design, acquire or construct, and operate one or more away-from-reactor storage facilities, and (2) accept for storage, until permanent disposal facilities are available, domestic spent fuel, and a limited amount of foreign spent fuel in cases when such action would further our non-proliferation policy objectives. All costs of storage, including the cost of locating, constructing and operating permanent geologic repositories, will be recovered through fees paid by utilities and other users of the services and will ultimately be borne by those who benefit from the activities generating the wastes.

Fourth, I have directed the Department of Energy to work jointly with states, other government agencies, industry and other organizations, and the public, in developing national plans to establish regional disposal sites for commercial low level waste. We must work together to resolve the serious near-term problem of low level waste disposal. While this task is not inherently difficult from the standpoint of safety, it requires better planning and coordination. I endorse the actions being taken by the Nation's governors to tackle this problem and direct the Secretary of Energy to work with them in support of their effort.

Fifth, the Federal programs for regulating radioactive waste storage, transportation and disposal are a crucial component of our efforts to ensure the health and safety of Americans. Although the existing authorities and structures are basically sound, improvements must be made in several areas. The current authority of the Nuclear Regulatory Commission to license the disposal of high level waste and low level waste in commercial facilities should be extended to include spent fuel storage, and disposal of transuranic waste and non-defense low level waste in any new government facilities. I am directing the Environmental Protection Agency to consult with the Nuclear Regulatory Commission to resolve issues of overlapping jurisdiction and phasing of regulatory actions. They should also seek ways to speed up the promulgation of their safety regulations. I am also directing the Department of Transportation and the Environmental Protection Agency to improve both the efficiency of their regulatory activities and their relationships with other Federal agencies and state and local governments.

Sixth, it is essential that all aspects of the waste management program be conducted with the fullest possible disclosure to and participation by the public and the technical community. I am directing the departments and agencies to develop and improve mechanisms to ensure such participation and public involvement consistent with the need to protect national security information. The waste management program will be carried out in full compliance with the National Environmental Policy Act.

Seventh, because nuclear waste management is a problem shared by many other countries and decisions on waste management alternatives have nuclear proliferation implications. I will continue to encourage and support bilateral and multilateral efforts which advance both our technical capabilities and our understanding of spent fuel and waste management options, which are consistent with our non-proliferation policy.

In its role as lead agency for the management and disposal of radioactive wastes and with cooperation of the other relevant Federal agencies, the Department of Energy is preparing a detailed National Plan for Nuclear Waste Management to implement these policy guidelines and the other recommendations of the IRG. This Plan will provide a clear road map for all parties and will give the public an opportunity to review the entirety of our program. It will include specific program goals and milestones for all aspects of nuclear waste management. A draft of the comprehensive National Plan will be distributed by the Secretary of Energy later this year for public and Congressional review. The State Planning Council will be directly involved in the development of this plan.

The Nuclear Regulatory Commission now has underway an important proceeding to provide the Nation with its judgment on whether or not it has confidence that radioactive wastes produced by nuclear power reactors can and will be disposed of safely. I urge that the Nuclear Regulatory Commission do so in a thorough and timely manner and that it provide a full opportunity for public, technical and government agency participation.

Over the past two years as I have reviewed various aspects of the radioactive waste problem, the complexities and difficulties of the issues have become evident—both from a technical and, more importantly, from an institutional and political perspective. However, based on the technical conclusions reached by the IRG, I am persuaded that the capability now exists to characterize and evaluate a number of geologic environments for use as repositories built with conventional mining technology. We have already made substantial progress and changes in our programs. With this comprehensive policy and its implementation through the FY 1981 budget and other actions, we will complete the task of reorienting our efforts in the right direction. Many citizens know and all must understand that this problem will be with us for many years. We must proceed steadily and with determination to resolve the remaining technical issues while ensuring full public participation and maintaining the full cooperation of all levels of government. We will act surely and without delay, but we will not compromise our technical or scientific standards out of haste. I look forward to working with the Congress and the states to implement this policy and build public confidence in the ability of the government to do what is required in this area to protect the health and safety of our citizens.

JIMMY CARTER

*   *   *   *

OFFICE OF THE WHITE HOUSE PRESS SECRETARY

THE WHITE HOUSE

FACT SHEET
THE PRESIDENT'S PROGRAM ON RADIOACTIVE WASTE MANAGEMENT

Highlights

In a Message sent to Congress today, the President outlined a comprehensive national radioactive waste management program. This program is based on the report of the Interagency Review Group on Nuclear Waste Management published in March, 1979.

The paramount objective in managing nuclear wastes is to protect the health and safety of all Americans, both now and in the future. The disposal of nuclear waste should not and will not be deferred to future generations.

The key elements of the President's program are:

- All levels of government share the responsibility for safe management and disposal of nuclear wastes.

  — In order to provide a more effective role for State and local governments the President has created a State Planning Council of elected State, local, and tribal officials and heads of cabinet departments and other federal agencies. Governor Richard Riley of South Carolina will serve as Chairman. State Representative Paul Hess of Kansas will serve as Vice Chairman. The Council will advise the Executive Branch and work with Congress on key radioactive waste management and disposal issues, especially related institutional decisions.

  — The basis of the relationship between States and the Federal government in the siting of high level waste repositories will be the principle of consultation and concurrence.

- Pending reviews required by the National Environmental Policy Act, an interim planning strategy for disposal of high level and transuranic waste has been adopted that relies on *mined geologic repositories.*

  — The program directed toward siting and opening repositories will be technically conservative, include expanded and technically diversified research and development and site investigations, and move carefully, in a step by step manner, toward site selection and operation of the first high level waste repository.

  — Immediate attention will focus on locating and characterizing a number of potential repository sites in a variety of different geologic environments with diverse rock types. This effort will be supported by a comprehensive research and development program. When four to five sites have been evaluated and found potentially suitable for a repository, one or more will be selected for further development as a licensed, full-scale repository. The site for the first full-scale repository should be selected by about 1985 and it should be operational by the mid-1990's.

— The Waste Isolation Pilot Plant (WIPP) project will be cancelled since it is unlicensed and cannot accept commercial wastes. The site of the proposed project at Carlsbad, New Mexico will be investigated further and if found qualified will be reserved for consideration along with other candidate sites in different geologic environments as a licensed repository for high level wastes.

• The safe interim storage of commercial spent fuel from nuclear power reactors will continue to be the responsibility of the utilities operating these plants until a permanent geologic repository capability exists. However, the Administration will continue to press for legislation to build or acquire limited spent fuel storage capacity at one or more away-from-reactor (AFR) facilities for those utilities unable to expand their storage capabilities and for limited amounts of foreign spent fuel when the objectives of the U.S. nonproliferation policy would be furthered.

• The Department of Energy will work with the States in their efforts to establish a reliable commercial low level radioactive waste disposal system.

• The Administration will submit legislation to extend Nuclear Regulatory Commission licensing authority to cover all DOE facilities for transuranic waste disposal and any new DOE sites for disposal of commercial low level waste. Under existing law, NRC has licensing authority over DOE facilities for disposal of high level radioactive wastes.

• Specific actions will be taken to improve and expedite regulatory actions by the Environmental Protection Agency and the Nuclear Regulatory Commission.

• The Nuclear Regulatory Commission is determining whether or not it has confidence that radioactive wastes can be disposed of safely. The President is urging the NRC to conduct its proceeding in a timely manner and to provide full opportunity for public, technical and government agency participation.

• The President's Fiscal Year 1981 budget for the Department of Energy requests $670 million in budget authority for nuclear waste programs. Other Department and agency requests total $49 million.

A brief description of the various types of nuclear waste and the quantities buried, stored and now being produced will be found in the background section of this Fact Sheet.

Objectives

The primary objective for waste management planning and implementation will be that existing and future radioactive waste from military and civilian activities (including commercial spent fuel if and when it is to be discarded) should be isolated from the biosphere and pose no significant threat to public health and safety.

The following principles will guide our program:

- The technical program must meet all of the relevant radiological protection criteria, as well as any other aplicable regulatory requirements. Although zero release of radionuclides or zero risk from any such release cannot be assured, such risks should fall within pre-established standards and, beyond that, be reduced to the lowest level practicable.

- The responsibility for establishing a nuclear waste management program will not be deferred to future generations.

- The nuclear waste management program should explicitly include consideration of *all* aspects of the waste management system including safety, environmental, organizational, and institutional factors.

- The basic elements of the program should be independent of the size of the nuclear industry and of the resolution of specific fuel-cycle or reactor-design issues of the nuclear power industry.

Elements of the President's Program

1. Relations with State and Local Governments

- The President has created, by Executive Order, a *State Planning Council* to advise the Executive Branch and work with the Congress in making and implementing decisions on waste management and disposal.

    The council will be chaired by Governor Richard Riley. There will be 14 members who are designated by the President as follows: eight governors; five state and local government officials other than governors; and, a tribal government representative. The Secretaries of Energy, Interior, and Transportation and the Administrator of the Environmental Protection Agency are also members.

    The Council will provide advice and recommendations to the President and the Secretary of Energy on nuclear waste management including interim storage of spent fuel. In particular, the Council will:

    (a) Recommend procedural mechanisms for reviewing specific nuclear waste management plans and programs, including the consultation and concurrence process designed to achieve Federal, State, and local agreement which accommodates the interests of all the parties.

    (b) Work on development of detailed nuclear waste management plans and provide recommendations to ensure that they adequately address the needs of affected States and local areas.

    (c) Advise on all aspects of siting and licensing of facilities for storage and disposal of nuclear wastes.

    (d) Advise on proposed Federal regulations, standards, and criteria related to nuclear waste management programs.

    (e) Identify and make recommendations on other matters related to the transportation, storage, and disposal of nuclear wastes that the Council believes are important.

- The principle of consultation and concurrence will apply in the siting of high level waste repositories. Under the framework of consultation and concurrence, a host State will have a continuing role with regard to the Federal government's actions on the siting, design and construction of a high level waste repository.

2. Interim Planning Strategy for High Level and Transuranic Waste Disposal

Pending reviews required by the National Environmental Policy Act and in order to provide interim guidance to the radioactive waste management program for its near-term actions and following the consideration of alternative technical approaches, the President has adopted a comprehensive interim planning strategy. The main components of the strategy are:

- Mined geologic repositories will be the primary focus of work for safe disposal of high level radioactive waste, including unreprocessed commercial spent fuel.

- The repository program will proceed in a technically conservative step-by-step manner, from the needed technical evaluations, through site selection, independent licensing review and ultimately to opening and operating a repository.

- Immediate attention will focus on (1) research and development, both in laboratories and at sites where underground workings can be used to study rock and waste from properties and interactions, and (2) locating and characterizing potential repository sites in different geologic environments and relying on diverse rock types.

- Once four or five sites in a variety of geologic environments have been evaluated and found potentially suitable for a repository, one or more will be selected for further intensive study or characterization and development as a licensed repository.

- Ultimately, several high level waste repositories will be opened, sited regionally insofar as technical considerations related to public health and safety permit.

Prior to proceeding with the first full-scale repository, an intermediate step might be taken by disposal of a relatively small quantity of high level waste in a licensed geologic test facility in order to gain experience applicable to subsequent actions with respect to full-scale repositories. Such a facility is not an essential component of a program leading to a full-scale, high level waste repository. It would provide an option, however, to test technical readiness and to exercise elements of the licensing process after an adequate site characterization program has been completed.

Following completion of environmental reviews required by the National Environmental Policy Act, the President will reexamine this interim strategy and decide whether any changes need to be made.

Following this strategy, the choice of site for the first full-scale repository should be made about 1985 and operations should begin by the mid-1990's. These dates reflect current estimates of the minimum time required to do the work necessary, including time for licensing and to permit full State and local government and public participation in decisionmaking.

The President's interim waste disposal strategy offers three important advantages:

(1) it provides maximum redundancy and conservatism so that no single or small number of setbacks would undermine the entire program, or even cause great delay;

(2) sites can be selected by comparing several locations among themselves thus providing greater confidence that the wastes will be disposed of safely;

(3) time will be available to put in place a good scientific program, to build procedures for licensing, public review and interaction, and to establish decisionmaking processes with State and local governments.

The Department of Energy is taking the following actions to implement this strategy:

• Regional, area and site investigations are being planned on a national basis to identify suitable high level waste repository sites. A variety of geologic environments and potential host rock types are being examined and this program will be expanded to ensure that the necessary sites will be available from which to select the first repository site.

• Research and development in laboratories and at test sites has been increased. Greater attention is now being given to a variety of possible waste forms, including spent fuel, to waste packaging and to waste-rock interactions under repository conditions.

• Three test facilities are planned: a granite facility in Nevada, a basalt facility in Washington State, and a salt facility at a site yet to be chosen.

The President has decided that the Waste Isolation Pilot Plant (WIPP) project as currently authorized will be cancelled. This project, for which construction has not yet commenced, is currently authorized for the unlicensed disposal of transuranic waste from our National defense program and for R&D using high level defense waste. Reasons for the cancellation are:

• Proceeding now on the basis of a single site is inconsistent with the strategy to compare sites with differing geologic characteristics prior to selection.

• An unlicensed facility is contrary to the President's policy.

• A facility for transuranic waste alone would provide no useful experience relevant either to licensing or to disposal of high level waste.

• It would also be an inefficient use of funds.

The site near Carlsbad, New Mexico which was being considered for this project will continue to be evaluated and, if qualified, will be reserved, along with other sites, for possible future use as a licensed repository for high level wastes. The DOE's FY1981 budget contains funds in the commercial nuclear waste program for protection of the Carlsbad site and continued characterization activities to determine suitability as a high level waste repository.

Although mined geologic repositories will be the focal point of the comprehensive national radioactive waste management program the DOE will continue to support a limited program directed toward other disposal alternatives. These include disposition of high level wastes in very deep boreholes and emplacement in ocean sediments in regions where the ocean floor is known to be geologically stable. These alternatives are considered to be longer range options to the mined geologic repository strategy.

3. Interim Storage of Defense and Commercial High Level Wastes.
The following actions are being taken to ensure safe and adequate care of defense and commercial nuclear waste in the interim period before a disposal facility is available:

- Adequate technical and financial resources will be made available to maintain defense wastes safely.

- Research and development at various defense facilities will proceed leading to plans for processing, packaging, and ultimate transfer to permanent repositories of transuranic and high level wastes from defense programs.

- Although spent fuel storage capacity is not an alternative to a permanent disposal capability, adequate storage must be provided until repositories are available. Primary responsibility for safe storage of commercial spent fuel lies with the utility industry. However, a limited amount of government storage capacity for commercial spent fuel would be desirable to provide flexibility to the national waste disposal program and an alternative for those utilities unable to expand their storage capabilities. The President continues to seek early enactment of legislation that would authorize the Department of Energy to: (1) design, acquire or construct, and operate one or more away-from-reactor storage facilities, and (2) accept for storage, until permanent disposal facilities are available, domestic spent fuel, and a limited amount of foreign spent fuel in cases when such action would further the objectives of our nonproliferation policy. Cost of constructing and operating away-from-reactor storage facilities will be borne by the users.

The Administration will continue to pursue both international and regional cooperative efforts to study nuclear fuel management options consistent with our nuclear nonproliferation policy.

4. Low Level Waste.
Three commercial burial grounds—South Carolina, Washington State and Nevada—are currently available to receive low level wastes originating in non-governmental industrial, medical and commercial activities. These have been operating intermittently because of inadequacies in waste packaging and shipment. Pressures have been mounting to find additional disposal sites to provide needed capacity for an ever-increasing inventory of commercial low level wastes.

The Department of Energy will work with the States to assist in their activities to establish regional disposal sites for low level wastes from the Nation's hospitals, research institutions, industry, and utilities. Until such time as additional disposal facilities can be sited and licensed, DOE and NRC will assist States in setting up interim storage facilities within the States. The State Planning Council will give low level waste management early, priority attention.

Other actions being taken in the area of low level waste include:

- DOE will review by 1981 alternative low level waste disposal techniques and determine whether any changes should be adopted in the future.

- DOE will accelerate R&D on improved methods of disposing of low level wastes.

- DOE will continue the existing land burial technology program presently designed to upgrade all DOE low level waste disposal operations by 1988.

- DOE will take action to ensure that adequate attention is given to the hydrologic characteristics of proposed locations for the future siting of low level waste disposal facilities.

5. Uranium Mill Tailings.

Past control of mill sites has been poor, with little or no attention to the problem of proper disposal of tailings upon completion of milling operations. The Uranium Mill Tailings Radiation Control Act of 1978 was passed to change this situation. EPA is directed to issue standards and criteria for disposal of mill tailings. NRC has licensing authority over active sites, and DOE is authorized to take remedial action at inactive sites.

The NRC and EPA are now developing standards, criteria, and regulations defiing acceptable levels of random emissions, siting, impacts on groundwater. The final Generic Environmental Impact Statement, (NUREG 0511, Draft issued 4/79) on uranium milling by NRC is nearing completion.

DOE, in cooperation with State governments, is now taking remedial action on abandoned tailings piles. DOE will continue to develop improved means of disposing of or stabilizing mill tailings over the long term.

6. Decontamination and Decommissioning.

As a general rule, unrestricted use of land will be the ultimate objective of D&D and institutional controls should not be relied upon after some period of time to provide long-term protection of people and the environment. However, because certain existing sites and/or facilities cannot be decontaminated at a reasonable cost, or perhaps at any cost, long-term institutional control may be required in these exceptional cases. These will require development of site-specific programs by NRC and DOE.

The following actions will be taken:

- DOE will prepare a nationwide plan for the decontamination and decommissioning of surplus facilities owned by DOE and other government agencies.

- DOE will work on designs for the construction of new facilities which will facilitate their eventual decommissioning.

- For new Federal facilities, decontamination and decommissioning specifications will be included in the initial design, and institutional arrangements will be made to ensure sufficient funding. The funding for D&D of government-owned facilities and sites will be through Federal appropriations. Responsibility and methods for financing D&D of licensed facilities will be determined by the regulatory process.

7. Transportation

Safe and reliable transportation of nuclear wastes is an essential component of the total waste management system. While complete assurance that release of radioactive material will not occur during normal operation or in serious accidents is impossible, it has been demonstrated that it is unlikely that a significant release can occur under most credible accident environments.

To improve the existing transportation system and enhance public confidence in it, the following actions are being taken:

- DOE is pursuing a program for testing and evaluating the performance of current and future generation waste packaging systems.

- The Department of Transportation is being directed to increase its management attention to nuclear waste matters and is completing its rulemaking on the role of Federal and local government bodies in routing of nuclear waste transportation along highways.

- DOT and NRC are working closely with the States to strengthen the nation's overall capability to respond to any transportation emergencies involving shipment of radioactive wastes.

- DOT will develop a data bank on shipment statistics and accident experience to be operational by 1982.

8. Financing.

The principle that will be applied to financing the cost of nuclear waste management and disposal is that the cost should be payed by the generator of the waste and borne by the beneficiary of the activity generating the waste. Utilities will pay the cost of storage and disposal of waste from power plant operations and pass these costs on to their customers. The government will pay the cost of storage and disposal of wastes from defense and government R&D activities and finance it from tax revenues.

9.   Regulatory Actions.

The Federal programs for regulating radioactive waste storage, transportation, and disposal are a crucial component of our efforts to ensure the health and safety. The following improvements are needed in the regulation of radioactive waste disposal:

- The current authority of the Nuclear Regulatory Commission to license the disposal of high level waste and low level waste in commercial facilities should be extended to also include the storage of spent fuel, as well as disposal of transuranic waste and non-defense low level waste in any new government facilities that might be built.

- The Environmental Protection Agency is responsible for creating general criteria and numerical standards applicable to nuclear waste management activities. The President has directed EPA to accelerate its schedule for the preparation of these criteria and standards and to prepare a positon paper that will indicate EPA's approach to setting standards and address the relationship between EPA's standards and actions taken by NRC and DOE.

- EPA and NRC will complete a Memorandum of Understanding dealing with coordinating methodologies and procedures.

- The Department of Transportation is responsible for regulating the transport of radioactive wastes, in part sharing that responsibility with NRC. The coordination between the two agencies is provided by an existing memorandum of Understanding between them. DOT is completing its rulemaking on the role of Federal and local government bodies in routing of nuclear waste transportation along highways.

- The Nuclear Regulatory Commission is now commencing a formal proceeding to determine whether or not it has confidence that radioactive wastes produced by nuclear power reactors can and will be disposed of safely. The President has urged the NRC to conduct this proceeding in a timely and thorough manner and to provide full opportunity for public, technical and government agency participation.

10. Legislation.
Legislation addressing the following areas will be submitted to Congress to implement the President's program.

- *State Planning Council.* To provide a permanent basis for the State Planning Council, which has been created by Executive Order.

- *Licensing Extension.* To implement the extension of NRC licensing authority to all new transuranic and new non-defense low level waste disposal facilities and any other facilities decided upon following the review of NRC's licensing study (NUREG 0527, September 1979).

- *Low Level Waste.* To assist the State in managing commercial low-level waste. The legislation will include authority for the States to enter into regional organizations or compacts for operation of the sites.

- *Decommissioned Facilities Surveillance.* To establish institutional responsibilities for long-term surveillance of formerly utilized Federal facilities which have been decommissioned and sold or otherwise released to unrestricted use.

The President will continue to press for authority to construct one or more away-from-reactor interim storage facilities for commercial spent fuel. This bill is already under consideration by the Congress. This legislation, or additional legislation, will implement the principle that costs of nuclear wastes disposal will be paid by the generator and borne by the beneficiary of the activity generating the waste.

11. Implementation and Management Structure.
Many Federal departments and agencies are necessarily involved in one or more aspects of nuclear waste management. In addition, the President's policy calls for full involvement of State, regional and local governments and organizations in program planning and execution. The President has designated the Secretary of Energy to be responsible for overall program integration and to establish necessary coordination mechanisms.

The Secretary of Energy will assume the lead role for: (1) coordinating all Federal nonregulatory aspects of radioactive waste management; (2) working out effective relationships with regulatory bodies such as the Environmental Protection Agency and the Nuclear Regulatory Commission; and (3) developing strong and effective ties between the Federal Government and the States on all aspects of radioactive waste storage and disposal. Within the Department of Energy, day-to-day activities are under the direction of an Assistant Secretary for Nuclear Energy (ASNE) who reports to the Under Secretary and the Secretary. Under the ASNE the Office of the Nuclear Waste Management (ONWM) is responsible for executing policy and managing all aspects of the nuclear waste management program.

Regulatory responsibilities are by law assigned to the Environmental Protection Agency, the Nuclear Regulatory Commission and the Department of Transportation. The Department of Interior has authority over Federal lands that might be used for waste storage or disposal and has extensive geoscience expertise in the U.S. Geological Survey. An Interagency Working Committee has been established by the Department of Energy to coordinate and integrate associated activities of DOE, DOI, EPA, NRC, DOT, and State. The President has also instructed DOE and DOI to prepare a Memorandum of Understanding between them delineating areas of cooperation and mutual responsibility and creating procedures to ensure they work jointly and reinforce each other's activities.

The primary planning mechanism will be a comprehensive National Plan for Nuclear Waste Management. The President has directed that this be produced by 1981 and be updated biannually thereafter. It is to be submitted for public review in draft and in revised form to the public and the Congress. The plan will include:

- summaries of the status of knowledge relevant to disposal of high level, transuranic, and low level radioactive wastes and uranium mill tailings.

- multi-year program plans for (1) interim management of high level radioactive waste and spent fuel; (2) site qualification for geologic repositories, and (3) R&D in the earth sciences and waste form and containers for high level and transuranic waste disposal;

- plans for low level wastes;

- a plan for decontamination and decommissioning of surplus government facilities;

- a plan for remedial action at inactive mill tailings sites;

- an integrated NEPA plan, covering the NEPA activities of all relevant agencies;

- updated cost estimates for all proposed activities;

- proposals to improve intergovernmental decisionmaking and resolution of environmental, economic and social issues associated with radioactive waste storage, transportation, and disposal;

- specific program goals and milestones for developing necessary regulations.

The President has issued detailed instructions to all federal agencies to ensure that his program will be implemented.

12. Public Participation.

It is essential that all aspects of the waste management program be conducted with the full disclosure to and participation by the public and the technical community. The President has directed the departments and agencies to develop and improve mechanisms to ensure such participation and public involvement consistent with any need to protect national security information and to comply fully with the National Environmental Policy Act. This includes providing technical and financial assistance to permit informed public input to programs and decisions and to support nongovernment efforts to increase social and technical understanding and agreement on nuclear waste issues. Formal mechanisms for receiving the best scientific and technical advice available and regular input from the interested public will also be strengthened.

13. International Cooperation.

Because nuclear waste management is a problem shared by many other countries and because selection of waste management alternatives has nuclear proliferation implications, the President will continue to encourage and support cooperative bilateral and multilaterial efforts which advance both our technical capabilities and our understanding of spent fuel and waste management and which are consistent with U.S. nonproliferation policy.

14. Funding.

Current funding levels ($ millions) for the respective agencies and programs are as follows:

	1979		1980		1981	
	BA	BO	BA	BO	BA	BO
Department of Energy						
Defense Nuclear Waste						
D&D	1	1	2	2	6	6
Interim Waste Management	147	187	163	213	219	222
Long-Term Waste Mgmt.	69	57	83	79	116	128
Terminal Storage[1]	36	49	28	26	0	11
Transportation	3	2	5	4	7	7
Program Direction	1	1	2	2	2	2
Subtotal	257	297	283	326	350	376
Commercial Nuclear Waste						
Commercial Waste Management	189	159	184	187	244	240
Remedial Actions	22	20	35	29	53	54
Program Direction	1	1	1	1	2	2
Subtotal	212	180	220	217	299	296
Spent Fuel						
Domestic	7	3	14	17	16	17
International	3	2	3	3	4	4
Program Direction	1	1	1	1	1	1
Subtotal	11	6	18	21	21	22
Total	480	483	521	564	670	694

Current funding levels (cont.)

	1979		1980		1981	
	BA	BO	BA	BO	BA	BO
Other Federal Nuclear Waste Programs						
Nuclear Regulatory Commission	20	18	26	23	38	34
Dept. of Energy (Environ.)[2]	3	3	3	3	4	4
Dept. of Interior	6	6	6	6	6	6
Dept. of Transportation[3]	1	1	1	1	1	1
Subtotal	30	28	36	33	49	45
GRAND TOTAL	510	511	557	597	719	739

1. Includes all expenditures related to the Waste Isolation Pilot Plant (WIPP) proposal.
2. This supports the remedial actions work conducted in the DOE Commercial Nuclear Waste Program.
3. The funding level for all years is under $1 million.

Additional Background Information

1. The Interagency Review Group Process.

As part of his initial National Energy Plan, the President ordered a review of the DOE nuclear waste management program. A Department of Energy task force carried out the review and published a "Draft Report of Task Force for Review of Nuclear Waste Management, February 1978."

With that report as a starting point, the President established the Interagency Review Group on Radioactive Waste Management (IRG) on March 13, 1978, to formulate recommendations leading to the establishment of a National policy for managing the Nation's nuclear waste with support programs.

The IRG was chaired by the Secretary of Energy and composed of representatives from the Departments of State, Interior, Transportation and Commerce, National Aeronautics and Space Administration, Arms Control and Disarmament Agency, Environmental Protection Agency, Office of Management and Budget, Council on Environmental Quality, Office of Science and Technology Policy, the Domestic Policy Staff, and National Security Council. The Nuclear Regulatory Commission was represented by a nonvoting member.

The IRG attempted to obtain a broad range of views from Congress, State and local governments, Indian Nations, industry, the scientific and technical community, public interest and environmental groups and the public. They published for public review and comment a draft report of their findings in October 1978 and over 15,000 copies were distributed. Seven small meetings (representing various special interests) and three regional public meetings were conducted to elicit public comments. Over 3,500 individual comments were received from State governments, industry, academia, environmental groups, and the general public. These comments were reviewed and summarized in the final IRG report, which was revised based on the comments. The final report was issued in March 1979 and formed the basis for the recommendations made to the President and ultimately was the basis of the President's policy statement.[1]

1. The final report "Report to the President by the Interagency Review Group on Nuclear Waste Management," March 1971 (TID-29442), and the technical report "Subgroup Report on Alternative Technology Strategies for the Isolation of Nuclear Waste," October 1979 (TID 28818) are available from the National Technical Information Service, U.S. Department of Commerce.

2. Definiton of Types of Wastes.

Nuclear wastes are produced in many different forms by a variety of activities including research investigations, medical diagnostics and treatment, mining and processing of uranium ore, defense-related nuclear activities and operation of commercial nuclear power plants. These wastes exist as gases, liquids and solids. The potential hazard of these wastes results from the fact that exposure to and/or uptake of the material can cause biological damage.

The major types of nuclear wastes are:

- *High Level Wastes*—These wastes are either fuel assemblies that are discarded after having served their useful life in a nuclear reactor (spent fuel) or the portion of the wastes generated in the reprocessing that contain virtually all of the fission products and most of the actinides not separated out during reprocessing.[2] These wastes are being considered for disposal in geologic repositories or by other technical options designed to provide long-term isolation of the wastes from the biosphere.

- *Transuranic Wastes*—These wastes are produced primarily from the reprocessing of defense spent reactor fuels, the fabrication of plutonium to produce nuclear weapons and, if it should occur, plutonium fuel fabrication for use in nuclear power reactors. Transuranic wastes contain low levels of radioactivity but varying amounts of long-lived elements above uranium in the Periodic Tables of Elements, mainly plutonium. This waste is currently defined as material containing more than 10 nanocuries of transuranic activity per gram of material.

- *Low Level Wastes*—These wastes contain less than ten nanocuries of transuranic contaminants per gram of material, or they may be free of transuranic contaminants. Although these wastes require little or no shielding, they have low, but potentially hazardous, concentration of quantities of radionuclides and do require management. Low level wastes are generated in almost all activities involving radioactive materials and are presently being disposed of by shallow land burial.

- *Uranium Mine and Mill Tailings*—These wastes are the residues from uranium mining and milling operations. They are hazardous because they contain low concentrations of radioactive materials which, although naturally occurring, contain long-lived radionuclides. The tailings, with a consistency similar to sand, are generated in large volumes—about 10 to 15 million tons annually—and are presently stored in waste piles at the site of mining and milling operations. A program is underway to either immobilize or bury these wastes to prevent them from being dispersed by wind or water erosion.

- *Decontamination and Decommissioning Wastes*—As defense and civilian reactors and other nuclear facilities reach the end of their productive lifetimes, parts of them will have to be handled as either high or low level wastes, and disposed of accordingly. Decontamination and decommissioning activities will generate significant quantities of wastes in the future.

- *Gaseous Effluents*—These wastes are produced in many defense and commercial

2. It is unclear whether the United States will reprocess commercial spent fuel. High level waste disposal facilities are therefore being designed to accept both spent fuel and waste from reprocessing.

nuclear activities, such as reactors, fuel fabrication facilities, uranium enrichment plans and weapons manufacturing facilities. They are released into the biosphere in a controlled manner, after passing through successive stages of filtration, and mixed with the atmosphere where they are diluted and dispersed.

3. U.S. Nuclear Waste Inventory and Forecast.

	Current (April 1979)		Annual Addition		Year 2000	
	DOE	Commercial	DOE	Commercial	DOE	Commercial
Disposed of (By volume— thousand cubic meters):						
Low Level Waste	1,470	515[1]	53	100	2500-6800	2800-7800[3]
Transuranic Waste	256	[2]	0	0	uncertain	[2]
Stored (By volume— thousand cubic meters):						
High Level Waste (Including spent fuel)	283	4.3	3	.6	320	40
Transuranic Waste	55	0	6	.3	250-3350[3]	6
Uranium Mill Tailings:						
Inactive sites (25)[4]	25	0	0	0	25	6
Active sites	0	125	0	15	0	425
Stored (By Radiactivity level—curies)						
High Level Waste (Including spent fuel)	$10^{9.5}$	4×10^7		$\sim 6 \times 10^{8}$[6]	10^9	$\sim 10^{10}$[6]
Transuranic Waste	Low	N/A	Low[5]	N/A	Low	N/A
Uranium Mill Tailings:						
Inactive sites	15,000	0	0	0	15,000	0
Active sites	0	$\sim 56,000$	0	6,800	0	$\sim 191,000$

1. As of 1/1/78.
2. Volume not available. Contains ∼125 kg TRU material at commercial disposal sites.
3. Range results from possible options on D&D of surplus facilities with waste quantity dependent upon mode of D&D for each facility.
4. Millions of Tons-Stabilization programs for inactive sites required by Uranium Mill Tailings Act of 1978.
5. Equilibrium exists. Annual additions equivalent to annual decay rate of ∼2×10^7 Ci.
6. Primarily spent fuel. Activity varies with age of material. Assumes average age of 10 years for additions and cumulative inventory. Activity @ 10 years = 1.05 × 10^5 Ci/MTHM.

• Present scientific and technological knowledge is adequate to identify potential repository sites for further investigation. No scientific or technical reason is known that would prevent identifying a site that is suitable for a repository provided that the systems view is utilized rigorously to evaluate the suitability of sites and designs, and in minimizing the influence of future human activities.

• A systems approach should be used to select the geologic environment, repository site, and waste form. A systems approach recognizes that, over thousands of years, the fate of radionuclides in a repository will be determined by the natural geologic environment, by the physical and chemical properties of the medium chosen for waste emplacement, by the waste form itself and other engineered barriers.

• The feasibility of safely disposing of high level waste in mined repositories can only be assessed on the basis of specific investigations at and determinations of suitability of particular sites.

• Some uncertainty about repository performance will always exist. Thus, in addition to technical evaluation, a societal judgment that considers the level of risk and the associated uncertainty will be necessary.

• Detailed studies of specific, potential repository sites in different geologic environments should begin immediately. Generic studies of geologic media or risk assessment analyses of hypothetical sites, while useful for site selection, are not sufficient for some aspects of repository design or for site suitability determination. Although most is known about the engineering aspects of a repository in salt, on purely technical grounds no particular geologic environment is an obvious preferred choice at this time.

• The actinide activity in transuranic wastes and high level wastes suggest that both waste types present problems of comparable magnitude for the very long term (i.e., greater than a thousand years).

• The degree of long-term isolation provided by a repository, viewed as a system, and the effects of changes in repository design, geology, climate, and human activities on the public health and safety can only be assessed through analytical modeling.

• Because it is not possible to predict or to restrict the activities of future generations, site selection guidelines, site suitability criteria, and repository design criteria must be developed in such a way as to minimize potential deleterious effects of human activities.

• Reprocessing is not required to ensure safe disposal in appropriately chosen geologic environments. Repositors can be designed to receive either solidified reprocessed waste or discarded spent fuel.

Attachment: Executive Order

EXECUTIVE ORDER

THE STATE PLANNING COUNCIL ON RADIOACTIVE WASTE MANAGEMENT

By the authority vested in me as President by the Constitution and laws of the United States of America, and in order to create, in accordance with the provisions of the Federal Advisory Committee Act, as amended (5 U.S.C. App. I), an advisory committee on radioactive waste management, it is hereby ordered as follows:

1-1. Establishment.

1-101.   There is established the State Planning Council on Radioactive Waste Management.

1-102.   The council shall be composed of eighteen members as follows:
(a) Fourteen members designated by the President as follows:
    (1) Eight Governors of the various states.
    (2) Five State and local elected government officials other than governors.
    (3) One tribal government representative.
(b) The heads of the following Executive agencies:
    (1) Department of the Interior.
    (2) Department of Transportation.
    (3) Department of Energy.
    (4) Environmental Protection Agency.
(c) The Chairman of the Nuclear Regulatory Commission is invited to participate in the activities of the Council; representatives of other departments and of United States territories and the Trust Territory of the Pacific Islands are invited to take part in the activities of the Council when matters affecting them are considered.

1-103.   The President shall designate a Chairman from among the members of the Council.

1-2. Functions.

1-201.   The Council shall provide advice and recommendations to the President and the Secretary of Energy on nuclear waste management (including intermin management of spent fuel). In particular, the Council shall:
(a) Recommend procedural mechanisms for reviewing nuclear waste management plans and programs in such a way to ensure timely and effective State and local involvement. Such mechanisms should include a consultation and concurrence process designed to achieve Federal, State, and local agreement which accommodates the interest of all the parties.
(b) Review the development of comprehensive nuclear waste management plans including planning activities for transportation, storage, and disposal of all catego-

ries of nuclear waste. Provide recommendations to ensure that these plans adequately address the needs of the State and local areas affected.

(c) Advise on all apsects of siting facilities for storage and disposal of nuclear wastes, including the review of recommended criteria for site selection and site suitability, guidelines for regional siting, and procedures for site characterization and selection.

(d) Advise on an appropriate role for State and local governments in the licensing process for nuclear waste repositories.

(e) Advise on proposed Federal regulations, standards, and criteria related to nuclear waste management programs.

(f) Identify and make recommendations on other matters related to the transportation, storage, and disposal of nuclear waste that the Council believes are important.

1-202.   Within one year after the Council's first organizational meeting, but in any event not later than seventeen months after the issuance of this Order, the Council shall prepare and submit to the President a public report on its functions set forth in Section 1-201.

1-3.   Administrative Provisions.

1-301.   Subcommittees of the Council may be established in accordance with the provisions of the Federal Advisory Committee Act, as amended.

1-302.   The members of the Council, including the members of its subcommittees, who are not otherwise paid a salary by the Federal Government, shall receive no compensation from the United States by virtue of their service on the Council, but all members may receive the transportation and travel expenses, including per diem in lieu of subsistence, authorized by law (5 U.S.C. 5702 and 5703).

1-303.   To the extent permitted by law, and subject to the availability of funds, the Secretary of Energy shall provide the Council, including any subcommittees, with necessary facilities, support, and services, including staff and an executive director.

1-4.   General Provisions.

1-401.   Notwithstanding the provisions of any other Executive order, the functions of the President under the Federal Advisory Committee Act, as amended (5 U.S.C. App. I), except that of reporting annually to the Congress, that are applicable to the Council, shall be performed by the Secretary of Energy in accordance with guidelines and procedures established by the Administrator of General Services.

1-402.   The Council shall terminate thirty days after it transmits its final report to the President, but in no event shall it terminate later than eighteen months after the effective date of this Order.

*   *   *

Appendix B: Policy Statement of the American Nuclear Society on High-Level Waste*

SUMMARY

The American Nuclear Society is concerned that the delay in constructing the first federal repository for nuclear waste disposal may impair the authorization of future nuclear power plants. Based upon various large-scale tests and engineering development in a 25-year research and development program, many prestigious studies in the past few years have concluded that safe disposal of nuclear wastes in a mined geologic repository is an acceptable approach.

Added confidence results from other considerations such as the two billion years of confirmatory evidence provided by the natural "reactor" in Gabon, Africa, and from the realization that the toxicity of nuclear waste gradually becomes less than that of naturally occurring materials that are routinely disposed of by modern society.

The American Nuclear Society believes that expeditious, forthright action is required on the part of the federal government to construct and place into operation a repository for permanent high-level waste disposal at the earliest possible time, as a part of a national nuclear waste disposal program.

INTRODUCTION

Radioactive waste disposal is an issue of crucial importance to the future of nuclear power. Assurance of public health and safety is of primary importance; this issue has broader implications, however, because in many instances the authorization of future nuclear electric power generation facilities has been made conditional upon a "demonstration" of safe, effective waste disposal. The American Nuclear Society is concerned

*Copyright 1979 by the American Nuclear Society.

242

about the potential impact that any delay in this connection might have on the nation's energy capability and about several misleading and/or inaccurate statements that have been issued regarding high-level waste disposal. The Society believes that, based on the current state of technology, a repository for permanent disposal can be designed, safely constructed, operated, and sealed. The Society further believes that this technology has been adequately developed through government-sponsored programs in the United States and abroad over the past 20 years; therefore, any associated moratorium on nuclear power is unfounded. In fact, any such action would represent an inexcusable denial to the American people of a valuable energy resource.

While the American Nuclear Society holds these beliefs, the realities of the current situation are that the general public will continue to be unduly concerned about the ultimate disposal of nuclear waste until such disposal is actually demonstrated. To alleviate these concerns, the American Nuclear Society believes that expeditious, forthright action is required on the part of the federal government to construct and place into operation a repository for permanent high-level waste disposal at the earliest possible time, as a part of a national nuclear waste disposal program.

The Society recognizes that additional data are required in specific technical areas and for specific sites and that a potential for improvement exists through development of advanced system concepts. These factors are common to any scientific/engineering development and particularly to those associated with the earth sciences. The Society believes that conservative design approaches can accommodate these uncertainties; further development, as described later, should be encouraged however. A discussion of key areas for further development and the basis for the Society's fundamental conclusions are given below.

BASIS FOR SAFETY

Consistent with recommendations of the National Academy of Sciences,[1] American Physical Society,[2] and the interagency review group on Nuclear Waste Management,[3] representing all federal agencies associated with energy production and environmental protection, the safe disposal of nuclear waste in a mined geologic repository is an acceptable approach. This disposal method is based on a series of engineered and naturally occurring barriers to the environment (biosphere), including:

- A high-integrity waste form with low leachability (e.g., borosilicate glass or crystalline materials).

- A canister within which the waste is placed.

- A backfill material, surrounding the canister, that could absorb or immobilize any waste that might leak.

- Extremely long transit times for any credible pathway from the repository deep below ground (on the order of tens or hundreds of thousands of years).

- A variety of minerals through which the waste must travel to reach the surface, any one of which might absorb the waste or chemically react and immobilize it.

• The further dilution of an already dilute waste product in any waterway that it might enter.

A repository can be engineered within current technology so that safe performance does not depend on any particular barrier and in most instances can be fully achieved by one barrier acting independently of the others. This is the same defense-in-depth concept that has resulted in the outstanding safety record of other nuclear facilities.

Several natural geologic formations, which are known to have been stable for many millions of years, are available for use as repositories. Some of these, such as bedded and domed salt formations, are also known to have been free from water intrusion over these same time periods, while others have had very low water intrusion. Since water transport is the only likely mechanism for waste removal, this is particularly encouraging. The past history of geologic formations is, to a great extent, *prima facie* evidence of safe performance. In order to emplace the waste, however, the formations would not be distrubed by man in two fundamental respects: creating potential pathways to the surface (i.e., shafts or boreholes) that must be sealed and emplacing the heat source of the waste itself. Both of these, in addition to uncertainties in geologic characterization, can be accommodated with current technology, through the use of appropriate design conservatisms and engineered barrier systems.

Shafts and boreholes will be constructed and sealed using standard mining and drilling techniques from the petroleum and related industries. These seals represent a vertical barrier of at least 2000 feet to the surface so that in the extremely unlikely event that some water were to leak in to the repository, it is even more unlikely that a mechanism exists to force the waste out.

The heat rate of emplaced waste can be controlled in a straightforward manner by several means, such as placing less waste in each canister, providing wider spacing between canisters, or allowing the waste to cool for a longer period before emplacement. Such techniques have been utilized in conceptual designs to maintain peak repository temperatures of less than 200°C, although some have contended that these temperatures are not low enough. If warranted, repository temperatures could be lowered even further, by means of these same techniques, to whatever value is deemed acceptable by regulatory authorites. Of course, these actions must be justified on the basis of cost-benefit evaluations.

The basis for safety in waste disposal is clearly established; although specific technical discussions continue, they do not affect this over-all conclusion. In fact, the public risk from radioactive waste disposal is substantially less than that from other toxic materials[4] that are routinely disposed of in modern society (e.g., arsenic and barium) and is several orders of magnitude less than risks we accept on a day-to-day basis (e.g., auto travel).[5]

DEMONSTRATION OF WASTE DISPOSAL CAPABILITY

Each phase of the technology necessary for safe disposal of radioactive waste has been demonstrated during the more than 20 years of federally sponsored research and development, and, through the adoption of conventional engineering and mining techniques for geologic disposal. Projected federal spending levels will assure application of these techniques as well as the development of more advanced methods. In addition, several other countries with substantial nuclear power commitments (including Canada, the United Kingdom, France, the Federal Republic of Germany, and Sweden) have independently developed similar disposal techniques and are proceeding with their implementation.

Apart from this compelling assurance of our ability to dispose of nuclear waste, moving forward with a national program for nuclear waste disposal is not predicated on future nuclear growth. The volume of waste that has already been generated through the U.S. nuclear defense effort is substantially greater than what is currently available from or is expected from civilian nuclear power for several years to come.[6] This defense-related waste has been handled and stored on an interim basis for more than 30 years without hazard to the public—this includes subsurface storage at several locations in the United States. Commercial nuclear waste is currently being stored within spent fuel at more than 40 power reactor locations around the country, awaiting ultimate disposition.

Numerous short-term demonstration programs and related research and development efforts have been carried out in support of commercial nuclear waste disposal. Spent fuel elements were successfully emplaced within a bedded salt formation near Lyons, Kans. for a two-year test period;[7] spent fuel assemblies are currently being emplaced in granite at the U.S. Department of Energy's Nevada Test Site;[8] and a Near Surface Test Facility is being constructed in volcanic basalt on the Hanford (Wash.) Reservation.[9]

Considerable experience has been derived from deep burial of intermediate-level commercial nuclear waste in salt formations with the Federal Republic of Germany.[10] But perhaps the most impressive demonstration is that of nature itself. In Gabon, Africa, a naturally occurring nuclear chain reaction almost two billion years ago created several tons of nuclear waste over tens of thousands of years.[11] Despite the fact that this waste was in an area of moving water and was not subject to any unusual geologic conditions, almost all plutonium and other transuranic elements and many of the fission products remained essentialy immobilized. Nature, therefore, has already supplied confirmatory evidence of the defense in depth provided by multiple naturally occurring barriers.

Although there are some who would advocate no action until the last technical detail is resolved or until a complete consensus is gained among all scientists (a practical impossibility for any major technical effort), the overwhelming body of technical evidence does not support this approach. Besides prior and ongoing demonstration programs, additional assurance will be afforded through extremely conservative designs and requirements to monitor waste performance during the 30-year repository filling time before a final approval for sealing is given. The American Nuclear Society concludes that future nuclear power growth should not be predicated on additional demonstration programs and that the cost/benefit of utilizing this valuable energy resource is readily justified.,

ADDITIONAL RESEARCH AND DEVELOPMENT

A considerable research effort is now under way in areas such as the performance of the waste form, the integrity of the barrier, the interaction between waste and rock, migration of radionuclides, thermo-mechanical effects on geologic formations, and modeling of risk and consequences. These programs will assist in resolving technical questions such as materials performance and the influence of thermal generation from waste. This work is important because a further understanding of these and similar subjects will likely allow further optimization of current conservative design approaches, resulting in a reduction in repository cost. It will also provide more definitive guidance for the relative comparison of geologic formations and potential sites within a given formation. Also, these programs will lead to more advanced disposal systems with improved design and performance. In the conduct of this work, there are several key areas that merit particular attention:

- *Site-specific investigations.* Although multiple natural barriers exist within a variety of stable formations and at a variety of sites within these formations, the relative performance of waste forms and other repository materials may vary with the specific geo-hydrologic and geochemical characteristics that are present. It is, therefore, advisable to characterize as many specific sites as is feasible and to model repository performance at these sites. This information would be very useful in scoping the national problem.

- *Consequence modeling/sensitivity.* A realistic modeling of potential release scenarios should be made to investigate consequences over the expected range of performance for key parameters. These investigations would identify which assumptions are of negligible importance to the over-all level of safety and, therefore, do not deserve further consideration. Conversely, they would identify those areas where continued work should be focused.

- *Waste-form performance.* Performance evaluations of waste forms alternative to borosilicate glass should be pursued as second-generation design, since these may offer the potential for improved performance over a broader range of temperatures and pressures. Evaluation should be performed, for example, on synthetic crystalline minerals, ceramic concretes, super-calcines, and cermets. Additional leach data for borosilicate glass under saturated and near-saturated conditions should also be developed for a variety of leachants.

- *High-integrity canisters.* The potential for use of very-high-integrity canisters, such as those contemplated in Sweden, should be investigated for potential domestic sites. These investigations should include a cost-benefit evaluation for implementation of such systems.

REFERENCES

1. National Academy of Sciences, Division of Earth Sciences, Committee on Waste Disposal, "The Disposal of Radioactive Waste on Land," NAS-NRC Publication 519 (1957); also, National Academy of Sciences, Committee on Radioactive Waste Disposal, "Geologic Criteria for Repositories for High-Level Radioactive Wastes" (August 3, 1978); both available from the National Technical Information Service, Springfield, Va. 22161.
2. Report to the American Physical Society on Nuclear Fuel Cycles and Waste Management Committee by the APS Study Group, *Reviews of Modern Physics* 50:1, Part II (January 1978).
3. Report to the President by the Interagency Review Group on Nuclear Waste Manaement, TID

29442 (March 1979); available from the National Technical Information Service, Springfield, Va. 22161.

4. B. Cohen, *Proceedings of the Conference on High-Level Solid Radioactive Waste Forms,* NUREG/CP-005, pp. 701, 777 (December 1978); available from the National Technical Information Service, Sprintfield, Va. 22161.

5. "Fatal and Injury Accident Rates on Federal-Aid and Other Highway Systems/1976," U.S. Department of Transportation (September 1978); available from the National Technical Information Service, Springfield, Va. 22161.

6. W. D. Rowe and W. F. Holcolm, "The Hidden Commitment of Nuclear Wastes," *Nuclear Technology* 24: 276-283 (December 1974).,

7. "Conceptual Design Report, Federal Repository, Lyons, Kansas," KE-NWTSR-71, Kaiser Engineers, Inc. (December 1971); available from the National Technical Information Service, Springfield, Va. 22161.

8. "NTS Terminal Waste Storage Program Plan for FY 1978," Nevada Operations Office, U.S. Department of Energy, Las Vegas, Nev.; available from the National Technical Information Service, Springfield, Va. 22161.

9. "Basalt Waste Isolation Program Annual Report—Fiscal Year 1978," RHO-BWI-78-100, Rockwell Hanford Operations, Richland, Wash. (October 1978); available from the National Technical Information Service, Springfield, Va 22161.

10. F. Gera and J. Olivier, "OECD Countries Pursue Geological Disposal," *Nuclear Engineering International,* pp. 35-57 (January 1978).

11. George A. Cowan, "A Natural Fission Reactor," *Scientific American* 235(1):36-46 (July 1976).

Appendix C: Executive Summary of the Interim Report of the State Planning Council

February 24, 1981

The President
The White House
Washington, DC 20500

Dear Mr. President:

The State Planning Council on Radioactive Waste Management was established under Executive Order No. 12192 (February 12, 1980) for the purpose of providing advice and recommendations to the President and the Secretary of Energy on Ways to strengthen the working relationship between the Federal government and state, local, and tribal governments in radioactive waste management.

The Executive Order called upon the Council to submit within twelve months of our first meeting, an Interim Report on its activities and accomplishments to date. On behalf of the Council, we therefore submit this Interim Report.

The State Planning Council came into existence as a direct result of the findings and recommendations of the Presidential Interagency Review Group on Nuclear Waste Management. The Council was developed in cooperation with and fully endorsed by all major public interest groups, including the national associations representing the Governor, state legislators, and local and tribal officials.

This Interim Report contains Council recommendations on consultation and concurrence for permanent repositories, low-level waste management, the National Plan for Radioactive Waste Management, transportation of radioactive waste, and a summary of the Council's deliberations on the interim storage of spent nuclear fuel.

As called for in the Executive Order, the Council will submit to you its Final Report on or before July 12, 1981.

Respectfully,

Richard W. Riley   Paul R. Hess
Chairman   Vice Chairman

Enclosures

EXECUTIVE SUMMARY

An effective radioactive waste management program in this country requires more than the solution of outstanding technical problems; it is equally dependent on the resolution of institutional issues. The importance of institutional issues was recognized by the President's Interagency Review Group on Nuclear Waste Management.

> The resolution of institutional issues required to permit the orderly development and effective implementation of a nuclear waste management program is equally important as the resolution of outstanding technical issues and problems.

In order to address this need, the State Planning Council on Radioactive Waste Management (SPC) was established by Executive Order of the President (No. 12192, February 12, 1980). That Executive Order charged the Council to provide advice and counsel to the President and to the Secretary of Energy on methods that would strengthen the working relationship between Federal, state, local, and tribal governments in radioactive waste management. In providing this advice and counsel, the SPC was asked to define policies and procedures which would help to resolve these institutional issues and to act as a vehicle for maintaining the intergovernmental dialogue on these issues.

The Council has identified general principles which should apply to all aspects of radioactive waste management and specific policies and procedures for implementing these principles. In formulating its recommendations, the Council has been instrumental in building an active dialogue among state, local, and tribal officials, among Federal agencies and non-Federal officials, and with industry and public interest groups.

THE PARTNERSHIP

Past radioactive waste management policies and procedures have often created an adversarial environment among Federal, state, and tribal officials. In an effort to create a decision making environment that is more conducive to finding solutions, the Council recommends a framework that will provide for an effective partnership among public officials. The principal elements of this framework are:

- a decision making process that rejects both Federal preemption and actions by states or tribes which arbitrarily prohibit their participation in the national radioactive waste management program;

• a working partnership that centers on interaction between the Federal government and state and tribal govenments, with the requirement that state and tribal governments establish procedures for participation by local officials, industry, and public interest groups, and the general public;

• the application of the partnership to all aspects of radioactive waste management and of waste regardless of the source; and,

• the full and timely exchange of information on plans and program activities.

The Council has identified various actions which should be taken to establish this framework, including:

• a national planning process, conducted through a structured dialogue on the National Plan on Radioactive Waste Management being developed by DOE, which will enable elected officials, the private sector, and the public to participate in establishing the goals and objectives of the national radioactive waste management program and the policies and procedures for achieving those goals and objectives;

• Federal legislation to define national policy for radioactive waste management;

• policy actions by the President and Cabinet secretaries to implement these policies;

• regulatory actions which can strengthen the partnership; and,

• actions by state, local, and tribal govenments which would insure an effective working partnership.

The following sections summarize Council recommendations for implementing this partnership in the following areas: high-level waste respositories, low-level waste management, the transportation of radioactive materials and waste, and the National Plan on radioactive waste management. There is also a summary of Council principles on the interim storage of spent nuclear fuel.

PERMANENT REPOSITORIES

The Council recommends that decisions on siting and development of permanent repositories be made through a process of consultation and concurrence. Consultation and concurrence involves neither a state/tribal veto nor a preemptive imposition of Federal will.

Consultation means the open and timely exchange of information among government officials in accordance with mutually agreed procedures. Concurrence is intended to be an incremental process, evolving from continuous consultation, and leading to a consensus among the public and their elected officials that the technical issues in and the social consequences of the repository development program have been appropriately analyzed and adequately addressed. The consultation and concurrence process should apply to all stages of the repository development effort.

The Council recommends that:

- consultation and concurrence consist of a process agreed upon by the parties involving a hierarchy of discussions and authorities, with the use of neutral third parties when appropriate, to resolve disagreements between the Federal government and a state or tribal government;

- a statutorily defined conflict resolution mechanism which calls upon the President or the Congress to make final siting decisions if the parties reach an impasse;

- a technical conservative, step-by-step repository development program based on the investigation of a number of sites. While recognizing the need to demonstrate the ability to dispose of high-level radioactive wastes, the Council has concluded that accelerated schedules will not provide sufficient time to build public confidence in proposed waste disposal solutions; and,

- the NRC licensing procedures and the NEPA process be kept in place as an integral part of the partnership process in radioactive waste management.

DOE together with various state and tribal governments are currently carrying out many of the elements of the consultation and concurrence process recommended by the Council. The Congress is urged to pass legislation to establish this framework as national policy.

LOW-LEVEL RADIOACTIVE WASTE

The council recommended that:

- the national policy on low-level waste be based on the fundamental principle that each state is responsible for the safe disposal of low-level waste generated by non-defense sources within its boundaries;

- states be encouraged to work together regionally to carry out this responsibility; and,

- the Federal government provide technical support and financial assistance to the states for low-level waste management.

The Congress adopted this framework for low-level waste management through the passage of the Low-Level Radioactive Waste Policy Act of 1980, which was signed into law by the President. This Act assigned responsibility for non-defense low-level waste disposal to the states and authorized them to form interstate compacts, subject to the approval of Congress, for the purpose of establishing and operating regional LLW disposal sites. Groups of states throughout the country have begun the process of region formation. The DOE and NRC are currently providing technical support and financial assistance to state and regional groups.

TRANSPORTATION

The Council recommends that:

- the Federal government develop, in cooperation with state and tribal governments, uniform national criteria for the highway routing of radioactive materials and waste based on public health and safety;

- each state and tribe be given the responsibility for making routing decisions within its jurisdiction, based on these national criteria and in cooperation with its local subdivisions;

- the Federal government work with state and tribal governments in the design of uniform prenotification criteria; and,

- state and tribal governments be responsible for emergency response.

The DOT has addressed the Council's recommendations on highway routing in its final rule on Highway Routing of Radioactive Materials (40 CFR Parts 171, 172, 173, 177). The DOT and NRC are currently addressing the issue of prenotification, and DOT and FEMA are working with states and tribes on mechanisms for emergency response. The Council is participating in these activities to assess the agencies' responses to Council recommendations.

INTERIM STORAGE OF SPENT FUEL

The Council has initiated its consideration of the interim storage of spent fuel. It has established the following principles as a basis for further discussion:

- The industry should be primarily responsible for providing storage capacity.

- If a Federal role in providing storage capacity is established, any Federal capacity should be limited to existing facilities for the storage of spent fuel from existing reactors, and such Federal facilities should be established regionally.

- Utilities should be required to maximize existing on-site storage.

- The construction of new reactors should include life-time storage capacity.

- When a disposal facility becomes available, the spent fuel should be systematically transferred from the Federal storage facility to the disposal facility.

NATIONAL PLANNING

The Council recommends that the national planning effort be carried forward under DOE to serve as a principal vehicle for establishing a partnership in radioactive waste management. An interactive process involving state, local, and tribal officials, the industry, public interest groups, and the general public can result in a truly "national plan" which represents a commitment to the safe handling and disposal of radioactive waste by all segments of our society. The Council also recommends that the national planning effort be carried forward in order to improve coordination among Federal departments and agencies and to provide stability and continuity to radioactive waste management programs.

DOE is currently developing a national plan, and the Council is working with the Department in conducting a review of the Fourth Working Draft by state and tribal governments in preparation for public review of the draft plan this summer.

THE COUNCIL'S ACTIVITIES

The council will submit to the President its Final Report, as called for in the Executive Order, by July 1981. The Council will continue its efforts to build an intergovernmental dialogue on radioactive waste management issues.

These efforts include developing networks to assist the Council in the evaluation of radioactive waste management issues, assisting individual states and tribes, and supporting regional groups on interstate compacts for low-level waste management. The Council will also maintain communications with the President, Federal Cabinet secretaries, and Congressional leaders, continue its cooperative efforts with national and regional associations of elected officials, and participate in meetings and conferences sponsored by national and regional associations and professional societies.

COUNCIL MEMBERS AND ALTERNATES

COUNCIL MEMBERS	ALTERNATES
CHAIRMAN *Richard W. Riley,* Governor State of South Carolina	*David M. Reid*
VICE-CHAIRMAN *Paul R. Hess,* Senator State of Kansas	*Elgie Holstein*
John N. Dalton, Governor State of Virginia	*William F. Gilley* *Jean Harris*
John V. Evans, Governor State of Idaho	*Robert Lenaghen* *Pat Costello*
Harry Hughes, Governor State of Maryland	*George Liebmann*
Ella T. Grasso,[1] Governor State of Connecticut	*Mary Hart*
Bruce King, Governor State of New Mexico	*George Goldstein*
Robert List, Governor State of Nevada	*Ralph DiSibio*
Dixy Lee Ray,[2] Governor State of Washington	*Betty McClelland*

1. Governor Grasso served on the Council until her resignation on December 31, 1980.
2. Governor Ray served on the Council until January, 1981, when her term expired.

Stanley Fink, Speaker of the Assembly State of New York	*Robert Kurtter*
Gordon Voss, Representative State of Minnesota	*Sara Meyer*
Ernest Morial, Mayor City of New Orleans	*Gino Carlucci*
Mary Louise Symon, Member Dane County Council Wisconsin	*Patrick Walsh* *Phyllis Dube*
Peter MacDonald, Chairman Navajo Tribal Council	*Robert Schryver* *Cathy Gramp*

The following Federal officials serve as members of the Council.

Secretary   *Joan Davenport*
Department of the Interior

Secretary   *Howard Dugoff*
Department of Transportation

Secretary   *Sheldon Meyers*
Department of Energy

Administrator   *David Rosenbaum*
Environmental Protection Agency

The following Federal official serves as an observer of the Council.

Chairman   *John Martin*
Nuclear Regulatory Commission

SPECIAL ADVISERS	
Jerome Joyce, Senator State of Illinois	*Richard Durbin*
William Wilkerson, Representative State of Mississippi	*John Green*

Appendix D: Participants of the Aspen Institute Conference November 15-17, 1979

Conference on the Governance Issues of Nuclear Waste Siting, Aspen Institute Program in Science, Technology and Humanism and Energy Committee, held at the John F. Kennedy School of Government, Harvard University, Cambridge, MA*

CHAIRPERSON

Paul Doty, Director, Aspen Institute Program in Science, Technology and Humanism, and Professor, Harvard University, Cambridge, MA

PARTICIPANTS

John Ahearne, Commissioner, Nuclear Regulatory Commission, Washington, DC
David Bazelon, Chief Judge, US Court of Appeals, Washington, DC
Junior Bridge, Office of Technology Assessment, Washington, DC
John Brodeur, Administrative Assistant to Rep. Jim Santini of Nevada, Washington, DC
Harvey Brooks, Professor, Harvard University, Cambridge, MA
Harry Browne, Vice President, Bechtel National, Inc., San Francisco, CA
Irwin Bupp, Business School, Harvard University, Cambridge, MA
John Busterud, President, RESOLVE, Palo Alto, CA
Albert Carnesale, Professor, Kennedy School of Government, Harvard University, Cambridge, MA
William Colglazier, Associate Director, Aspen Institute Program in Science, Technology and Humanism, Kennedy School of Government, Harvard University, Cambridge, MA
Ben Cooper, Committee on Energy and Natural Resources, US Senate, Washington, DC

*Affiliations of participants are given at the time of the conference.

255

Thomas A. Cotton, Office of Technology Assessment, Washington, DC
Robert Craig, President, Keystone Center for Continuing Education, Keystone, CO
David Deese, Kennedy School of Government, Harvard University, Cambridge, MA
Lee Dembart, Los Angeles Times, Los Angeles, CA
John Driscoll, Kennedy School of Government, Harvard University, Cambridge, MA
John Gervers, Coordinator, WIPP Task Force, State Capitol, Santa Fe, NM
Ali Ghovanlou, Metrek Division, The Mitre Corporation, McLean, VA
John Gibbons, Director, Office of Technology Assessment, Washington, DC
Robert Gillette, Los Angeles Times, Los Angeles, CA
Keith Glaser, Senate Subcommittee on Nuclear Regulation, Senate Committee on
    Environment and Public Works, Washington, DC
William Gouse, Vice President, Metrek Division, The Mitre Corporation, McLean, VA
Harold Green, Esq., Fried, Frank, Harris, Shriver & Kampelman, Washington, DC
Ted Greenwood, Dept. of Political Science, Massachusetts Institute of Technology,
    Cambridge, MA
Michael Haltzel, Project Director, Aspen Institute Committee on Energy, President's
    Office, New York University, New York, NY
William Hogan, Kennedy School of Government, Harvard University, Cambridge, MA
Roger Kasperson, Director of CENTED, Clark University, Worcester, MA
Terry Lash, Natural Resources Defense Council, San Francisco, CA
Henry Lee, Research Program Coordinator, Energy and Environmental Policy Center,
    Kennedy School of Government, Harvard University, Cambridge, MA
Dan Luria, Research Dept., United Automobile Workers, Detroit, MI
Frank Marriott, Director of Strategic Programs, Westinghouse Electric Co.,
    Washington, DC
Jessica Tuchman Mathews, The Washington Post, Washington, DC
Tom Moss, Science Advisor to Rep. George Brown, Washington, DC
Jerry Murphy, Energy Research Group, Waltham, MA
Sheldon Myers, Program Director, Office of Nuclear Waste Management, Dept. of
    Energy, Washington, DC
Joseph Nye, Professor, Kennedy School of Government, Harvard University, Cam-
    bridge, MA
Larry O'Donnell, General Atomic Company, San Diego, CA
Dana Orwick, Special Consultant, Aspen Institute, Washington, DC
Dorothy K. Powers, Chairperson, Energy Committee, National League of Women
    Voters, Princeton, NJ
Marvin Resnikoff, Sierra Club, Buffalo, NY
Gene Rochlin, University of California at Berkeley, Berkeley, CA
Randy Rydell, Center for Science and International Affairs, Kennedy School of Gov-
    ernment, Harvard University, Cambridge, MA
Henry Schilling, Battelle Human Affairs Research Centers, Seattle, WA
Jean Schrag, Minority Staff, Committee on Energy and Natural Resources,
    Washington, DC
John Selby, Chairman of the Board and President, Consumers Power Company,
    Jackson, MI
Steven Sklar, Chairman, Nuclear Energy Subcommittee, National Conference of State
    Legislatures, Baltimore, MD

Richard Tobe, Executive Director, Legislative Commission on Science and Technology, Albany, NY

Emilio Varanini, Commissioner, California Energy Commission, Sacramento, CA

Michael Ward, Subcommittee on Energy and Power, House Interstate and Foreign Commerce Committee, Washington, DC

Len Weiss,. Staff Director, Subcommittee on Energy, Nuclear Proliferation and Federal Services, Governmental Affairs Committee, Washington, DC

Robert Williams, Electric Power Research Institute, Palo Alto, CA

Shirley Williams, Institute of Politics, Kennedy School of Government, Harvard University, Cambridge, MA

William Willis, General Manager, Tennessee Valley Authority, Knoxville, TN

Albert Wilson, Special Advisor to Gov. Evans and Dean of Engineering, Idaho State University, Pocatello, ID

Dorothy Zinberg, Kennedy School of Government, Harvard University, Cambridge, MA

Charles Zraket, Executive Vice President, The Mitre Corporation, Bedford, MA

Index

259

About The Editor and Contributors

E. William Colglazier, Jr., is a former research physicist who now works on public policy issues connected with science and technology. He is a Research Fellow at the Center for Science and International Affairs at Harvard University, and from 1978 to 1981 was also Associate Director of the Aspen Institute Program in Science, Technology and Humanism. He received his Ph.D. from the California Institute of Technology and has worked at the Stanford Linear Acceleration Center and the Institute for Advanced Study in Princeton. In 1976-1977, he was a Congressional Science Fellow sponsored by the American Association for the Advancement of Science.

David A. Deese is Assistant Professor of Political Science at Boston College and Research Associate at Harvard University. He was formerly Assistant to the Director of the Center for Science and International Affairs, Harvard University, where he directed the Energy and Security Project and other research programs. He has published numerous articles and three books: *Nuclear Power and Radioactive Waste* (Lexington Books, 1978), *Nuclear Proliferation: The Spent Fuel Problem,* edited with Frederick C. Williams, (Pergamon, 1979) and *Energy and Security,* edited with Joseph S. Nye (Ballinger with Harper & Row, 1981).

Paul Doty is Mallinckrodt Professor of Biochemistry and Director of the Center for Science and International Affairs at Harvard University. He is also a Senior Fellow of the Aspen Institute and Director of the Aspen Arms Control Consortium.

Harold P. Green is Professor Emeritus of Law at the George Washington University National Law Center and is a partner in Fried, Frank, Harris, Shriver & Kampelman in Washington. He has written extensively on energy and nuclear energy topics and is co-author of *Government of the Atom: the Integration of Powers* and author of the *Energy Law Guide* which is a part of the Callaghan Energy Law Service of which he is general editor.

Thomas H. Moss is Staff Director of the Subcommittee on Science, Research and Technology of the House Committee on Science and Technology. He received his Ph.D. in physics from Cornell and for eight years was a research staff member specializing in biophysics at the IBM Research Center in Yorktown, New York. He was a member of the American Physical Society's Nuclear Fuel Cycle and Waste Management Study of 1975-1977 and co-chairman of the New York Academy of Sciences' retrospective conference on the Three Mile Island accident. He has received a distinguished service award for teaching science courses to high school dropout students and has been a consulting editor of *Environment* magazine.

Marvin Resnikoff is a co-project-director and staff scientist for the Sierra Club Radioactive Waste Campaign. He also teaches a course on "energy for the future" at Rachel Carson College, State University of New York at Buffalo. His friends call him a political physicist.

Emilio E. Varanini, III, an attorney, played a leading role in establishing the California Energy Commission, to which he was appointed in 1976. He has participated in many national forums on radioactive waste disposal, including testimony before Congress, and has authored numerous papers and reports on various aspects of the waste disposal problem. Through his efforts, the California Energy Commission has provided support to the Swedish government in this area.

L. Mark Zell, a graduate of Princeton University and the University of Maryland School of Law, is a partner in the firm of Topf & Zell of Washington, D.C. and is a member of the bars of Washington, D.C. and Maryland.

Dorothy S. Zinberg is Lecturer in Public Policy and Director of Seminars and Special Projects at the Center for Science and International Affairs at the Kennedy School of Government at Harvard University. For ten years a biochemist at the Harvard Medical School, she later received a Ph.D. in sociology at Harvard. She is Chairperson of the Advisory Committee of the International Divison of the National Science Foundation, a member of the Council on Foreign Relations, and a member of the Commission for International Relations at the National Academy of Sciences. Her current research includes an evaluation of the social dimensions of the energy transition.

Ted Greenwood is Associate Professor of Political Science and is associated with the Center for International Studies at the Massachusetts Institute of Technology. He served as a senior policy analyst in the Office of Science and Technology Policy from 1977 to 1979 where his responsibilities included participation in the formation of Carter administration policy on radioactive waste management.